INTRODUCTION TO
EQUINE SCIENCE

Third Edition

Shannon Pratt-Phillips

North Carolina State University

Cover image © Shutterstock.com

Kendall Hunt
publishing company

www.kendallhunt.com

Send all inquiries to:
4050 Westmark Drive
Dubuque, IA 52004-1840

Contents

CHAPTER 1

Introduction to Horses

The objectives of this section are to become familiar with the basic parts of the horse (external anatomy) and terminology associated with horses

Parts of the Horse

- Fetlock
- Pastern
- Hoof
- Coronet band
- Barrel
- Withers
- Back
- Loin
- Croup
- Dock
- Tail
- Gaskin
- Flank
- Stifle
- Hock
- Muzzle
- Forehead
- Poll
- Throatlatch
- Cheek
- Forelock (hair)
- Mane (hair)
- Crest (of the neck)
- Shoulder
- Point of Shoulder
- Chest
- Elbow
- Forearm
- Knee
- Cannon

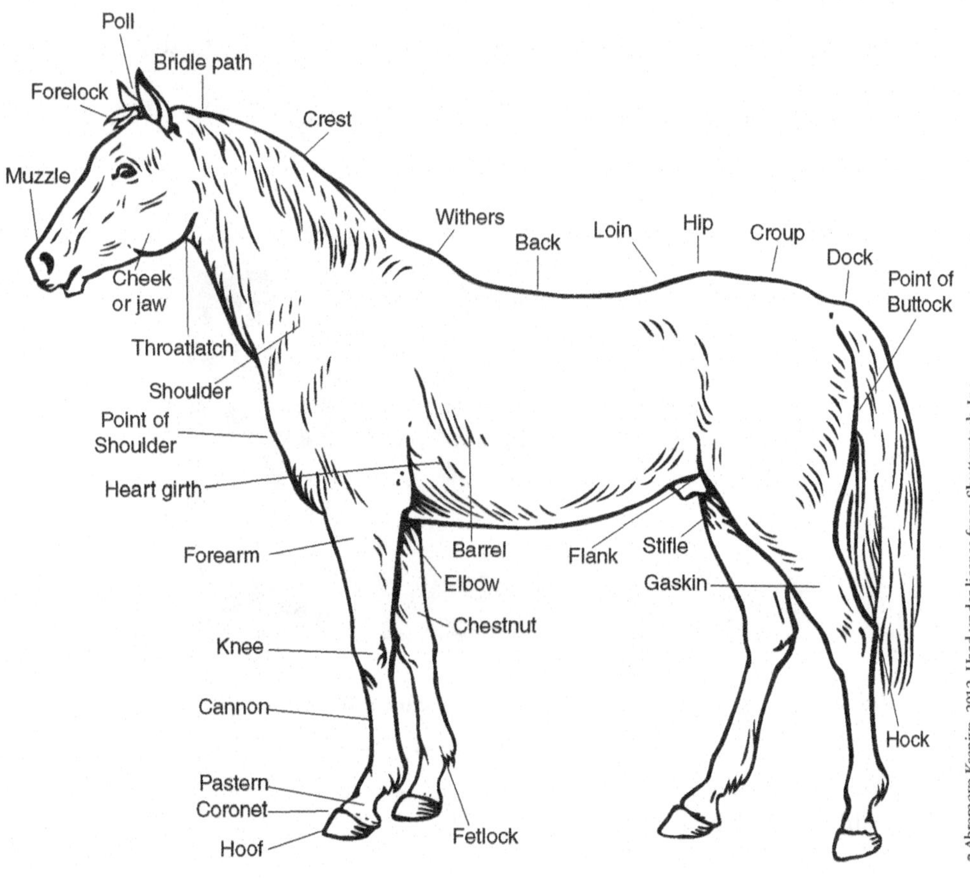

Additional notes about this picture:

The bridle path refers to the area where the bridle's crown- or headpiece would go. The mane and forelock junction in this area is often clipped to provide a "path" for the bridle.

The point of the shoulder should represent the point that the scapula (shoulder blade) makes, while the chest represents the front of the horse in between their front legs and their neck.

The point of the buttock is actually a little higher than the arrow indicates – and it is the furthest distal point of the horse's hind end.

The heart girth is actually the circumference of the horse's girth area, from behind the horse's elbow and up around just behind the withers.

The chestnut is a callous type of structure that is believed to be a vestigial toe pad. They are typically found on all limbs inside of the knee and the hock.

Basic Terminology

Horse: A hoofed mammal of the Equidae family (*Equus ferus caballus*)

Pony: A horse with small mature size

Breed: A group of horses with common ancestry and/or common characteristics

Feral horse: A wild (free ranging) horse that is a descendent of a domesticated horse

Hand: Term for measuring horse height. 1 hand = 4 inches

Mare: A mature female horse

Stallion: A mature male horse

Gelding: A mature male horse that has been castrated (or "gelded")

Dam: A horse's female parent

Sire: A horse's male parent

Foal: A newborn horse

© Rita_Kochmarjova/Shutterstock.com

Weanling: A young horse that has been weaned from its dam

Yearling: A horse that is one year of age (up to 2 years of age)

Filly: A young female horse

Colt: A young male horse

Gait: The order of how the horses' feet fall when the horse is in motion. Includes the walk, trot, canter and gallop plus other breed specific gaits.

Tack: A general term for the equipment used on the horse

Bridle: A piece of tack that is used for riding. Includes a *bit* that goes in the horse's mouth and *reins* that are held by a rider.

Saddle: A piece of tack that is placed on the horses back and where the rider sits for riding

Halter: A piece of tack that goes on the horse's head and is used for leading or tying

Lead shank (or lead rope): A rope or leather "leash" for leading horses

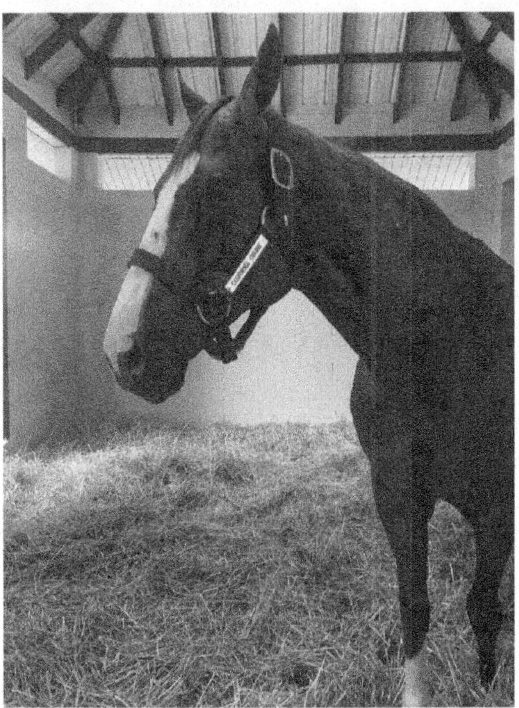

Source: Shannon Pratt-Phillips

Basic colors

These are the basic most common colors of horses. Several more will be discussed later in chapter 3.

A good overview and pictures can be found here:
https://en.wikipedia.org/wiki/Equine_coat_color

Black: A horse with black hair on its body and a black mane and tail

Bay: A horse with reddish or brown hair on its body and black "points" – its mane, tail and lower legs

Chestnut/Sorrel: A red or brown body color with brown (or red or flaxen) mane and tail

Palomino: A horse with a golden or yellow color on its body with a light mane and tail

Buckskin: A horse with a yellow or beige body color with black points

Gray: A horse with a mix of white and dark hairs over black skin. Horses "grey out" over time, similar to some people who go grey with age.

Dun: A horse with a yellow or beige body, dark mane and tail and "primitive" markings such as a dorsal stripe or zebra stripes on its legs

Comparative Anatomy

The horse's limbs are similar in basic anatomy to our own, though rather than having 5 digits, they have one. When looking at digits in your own hand, start with your thumb as "1" and your pinky is digit "5". In the horse, they walk around on digit 3.

Horse	Human
Knee (front limb)	Wrist
Cannon	Long bones in hand (or foot)
Fetlock	First knuckle
Hoof wall	Fingernail
Stifle	Knee
Hock	Heel/ankle

CHAPTER 2

History of Horses

The objective of this section is to introduce students to how the horse has developed throughout history, both in physical structure through the process of evolution, and in the horse's role in society through domestication.

Section 2.1 Evolution of the Horse

The first horse-like creature was quite unlike what we know today as the modern horse. In fact, throughout history, many horse-like creatures have existed, with seemingly progressive changes to our modern horse. However, such progression was not in a straight line, but rather different horse-like creatures existed and flourished at different times in history. Today's horse is simply the only one that survived. Many of the traits that existed in each equid species reflected the changes in the environment and landscape in which they lived. The trends that have occurred throughout history are not representative of successive changes from one equid species to the next, but rather are the result of an overall review of the equid species throughout history.

It is of interest to note that today's study of evolution includes not only the archeologic record of excavation sites, but also genetic testing of remnant DNA. This is an exciting area of genetics that is helping to reveal some very interesting ties between ancient horses and today's horses.

General Trends

There are four general trends that occurred throughout history and the evolution of the horse.

1) The body size increased. The equids of the past tended to be small, with the first recognized ancestor being the size of a fox.
2) The progression from 4 or 3 toes to a single hoof. The earliest horse-like creatures had toes that resembled the toes of a dog, with early hooves forming. They also had 5 digits (similar to our own fingers; counted 1-5 starting with the "thumb") with only 3 (front limbs) or 4 (hind limbs) of these digits touching the ground. Over time, the other digits regressed and became vestigial. Today's horse is a monodactyl (one toe; the middle 3rd digit), though digits 2 and 4 remain as the splint bones. These changes occurred as the environment became less swampy, and the single digit likely allowed horses to become faster runners.
3) The teeth developed to become suited to grazing. The early equids likely were browsers, and ate the more succulent parts of the plant such as the leaves. Therefore their teeth needn't be big and coarse and were in fact small and more similar to our own. As their living environments changed, equids started to eat more of the stalks of the plants and became grazers. This necessitated the need for their molars and pre-molars to be more suited to grinding to coarser fibers found within these types of plants. The jaw of the

horse also is somewhat unique compared to other grazing animals, in that they still retain upper incisors (unlike cattle) and their lower jaw rests just inside their upper jaw.

4) The skull elongated. This facilitated room for more teeth, and additional changes such as the position of the eyes. The skull remains lightweight thanks to large nasal passages.

Key Ancestors

The Equidae family includes today's horses, donkeys and zebras, as well as extinct species.

Hyracotherium (or Eohippus, which means "dawn horse") is the first recognized horse-like creature. It existed approximately 52 million years ago and was the size of a fox (about 14 inches tall). Hyracotherium had toes that were starting to resemble hooves, with 4 toes on each front foot and 3 toes on each hind foot. In the front foot, the thumb digit (numbered "1") was still present, but did not touch the ground, while the remaining four toes (2-5) did. Similarly in the hind foot, the digits 1 and 5 were considered vestigial. There have been many full skeletons of these horses found.

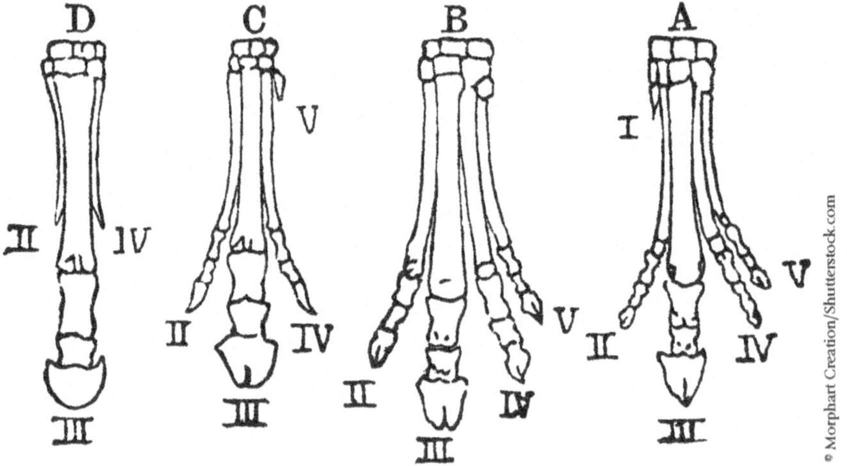

Foot of Hyracotherium

Skeleton of Hyracotherium

Hyracotherium

Mesohippus ("middle horse") lived about 40-35 million years ago in North America. It had three toes on each foot and the middle toe was beginning to dominate. It was taller than Hyracotherium at about 24 inches tall. Its teeth were becoming more suited to grazing the changing grasslands.

Miohippus ("lesser horse") lived about 36 million years ago and survived through into the Miocene era (24-5 million years ago), thus overlapping Mesohippus for some time. It was also larger than Hyracotherium, had further developed teeth and deep facial fossa (hollows).

Merychippus ("ruminant horse" – though not technically a ruminant) lived in North America about 20-10 million years ago. It had three toes on each foot and was approximately 35 inches tall. It was one of the first true grazers, and many other grazing horses are descendants from this equid, including Hipparion, Protohippus and Pliohippus.

*a "ruminant" is an animal that ruminates – or chews its regurgitated cud.

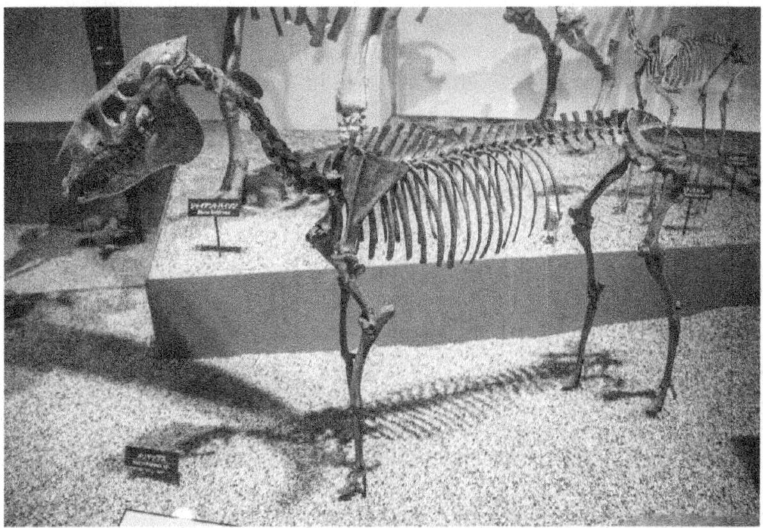

Skeleton of Merychippus

Hipparion was the size of a pony and a nose similar to a tapir. They lived in North America, and later migrated to Europe and Asia.

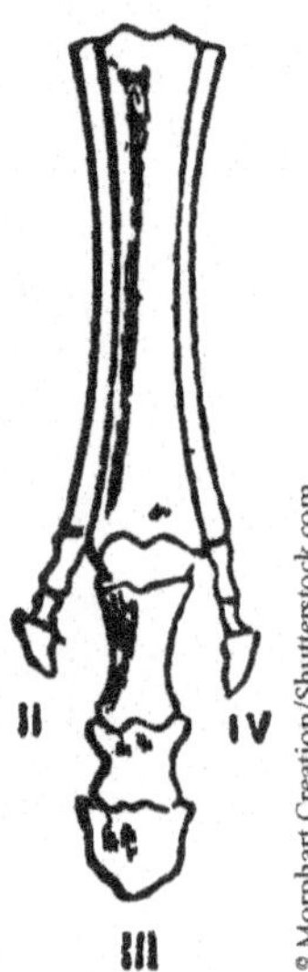

Foot of Hipparion

Pliohippus was considered the first true monodactyl (one toe) and lived around 12 million years ago.

Dinohippus lived in North America approximately 10 million years ago and is considered a close ancestor of Equus.

Equus is the genus in which all living equids belong to including horses, donkeys and zebras. One of the oldest Equus species is *Equus simplicidens* (Hagermans horse) and lived approximately 3.5 million years ago. It was about 50 inches tall and resembled a zebra in body type. It lived in North America (with the oldest fossils found in Idaho) and spread to Europe and Asia. Other Equus ancestors lived approximately 6 million years ago, with divergencies of the Asian hemiones and Afrian Zebras.

Genetic evidence has recently shown there were three equid species that diverged from a common ancestor. The Hippidion (little horse) is a genus from South America, and lived from 2 million years ago until about 11,000. The Haringtonhippus (stilt-legged horse; New World Stilt-Legged horse) lived North America from 4-5 million years ago to more recent times. The third equid species, equus, the true horse, includes other prehistoric horses, the Przewalski's horse and the Tarpan. These horses migrated to Europe via the Bearing straight, and once into Europe and Asia, they were domesticated.

All remaining species of equids went extinct in America approximately 8,000-12,000 years ago.

It is believed that equine subtypes gave way to different breeds as we know them today.

The Warmblood subspecies or Forest horse.

The Draft subspecies, characterized by a small, but heavy body and heavy coat adapated to colder climates.

The Oriental subspecies, that gave way to the Arabian and Akhal-Teke

Taxonomic Classification of Modern Horses

Kingdom: Animalia

Phylum: Chordata

Class: Mammalia

   Order: Perissodactyla

Family: Equidae

     (**Subfamily**: Equinae)

       **All other Equidae other than Equus are extinct

Genus: *Equus*

  Subgenus: Equus

    Equus caballus (Domestic Horse)

    Equus ferus (Tarpan, aka Eurasian wild horse)

    Equus przewalskii (Przewalski's horse, aka Mongolian Wild Horse)

      *sometimes called *Equus ferus caballus, Equus ferus ferus, Equus ferus przewalskii*

  Subgenus Asinus

    Equus africanus (African wild ass)

    Equus africanus asinus (Domestic Donkey)

    Equus hemionus (Onager)

    Equus hemionus khur (Khur)

    Equus kiang (Kiang)

© Alan Jeffery, 2011. Used under license from Shutterstock, Inc.

© Brendan Howard, 2011. Used under license from Shutterstock, Inc.

© Alan Jeffery, 2011. Used under license from Shutterstock, Inc.

© Alistair Michael Thomas, 2011. Used under license from Shutterstock, Inc.

© EcoPrint, 2011. Used under license from Shutterstock, Inc.

© Hector Arencibia, 2011. Used under license from Shutterstock, Inc.

Why did the horse survive? Likely thanks to domestication.

Subgenus Dolichohippus

Equus grevyi (Grevy's zebra)

Subgenus Hippotigris

Equus quagga (Plains zebra)

Equus quagga burchelli (Buchell's zebra)

Equus quagga beohmi (Grant's zebra)

Equus quagga quagga (Quagga, extinct 20th century)

Equus zebra (Mountain zebra)

Equus zebra zebra (Cape Mountain zebra)

Plus many other live and extinct animals…

Section 2.2 Domestication

Domestication refers to when humans have control over the breeding and productivity of animals.

There are six criteria for domestication as summarized by Jared Diamond, an evolutionary biologist.

1) They must be able to consume a **flexible diet**.
2) The animals must **mature quickly**.
3) They should be able to be **bred in captivity**.
4) The animals should have a **good disposition**.
5) They should be **unlikely to panic**.
6) The animals should **recognize humans as their leader** in the social hierarchy.

What other animals are domesticated?

Evidence for early domestication includes changes in the skeleton (changes in size) that suggest selective breeding has occurred. Changes in the teeth might suggest the use of a bit and are evidence they were worked. Other evidence includes changes in geographic distribution such that remains of horses were found in locations where they previously hadn't been. Archeological artifacts such as art, bits, weapons or even chariots are also evidence of domestication. Similarly, the presence of horses in corrals or gravesites also gives support to their roles in early society. Being able to use horses for transportation changed the lives of ancient people, by being able to increase mobility, warfare strength, communication and commerce and of course, agriculture.

Genetic evidence of domestication is associated with either the Y chromosome (paternal lines) or mitochondrial DNA (maternal lines). Rapid increases in coat color variation strongly suggests selective breeding.

Cave drawings in France dating to almost 20,000 BCE show evidence of horses being hunted and used for meat. Also in France is the Rock of Solutré, where thousands of horse skeletons have been found at the bottom of the cliff. It is believed that hunters chased horses up the hill and off the cliff, so they could eat the meat below.

There are a few archeological sites that give some evidence to some early domestication and use of the horse around 4000 BCS.

Dereivka is a site in the Ukraine on the Eurasian Steppes, where more than 50% of mammal bones that were identified belonged to horses. There is question as to how

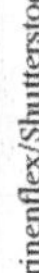

much these horses were truly domesticated, vs. intensively hunted (Levine, 1990, DOI:10.1017/S0003598X00078832)

The Botai people of Kazakhstan also have some evidence of early domestication of both horses and dogs. Evidence includes horse bones in garbage deposits hand indents on teeth could suggest bit use. Concentrated areas of horse dung and corrals also suggest their domestication by the Botai. However, there is also newer evidence to suggest that the horses use by the Botai were in fact the Przewalski's horse, and may have also merely been intensively hunted. (Taylor and Barrón-Ortiz, 2021; https://doi.org/10.1038/s41598-021-86832-9)

Thus, with domestication, the first uses of horses were likely meat.

It is becoming more likely that the first populations to use horses beyond meat were the Sintashta, of the Black Sea Steppe region, closer to 3000 BCE. Here, equine milk proteins have been found in human teeth, confirming the consumption of milk.

Only later were horses used for transportation. There is some argument as to whether or not horses were first ridden or driven. Bit artifacts and dental wear may support the use of bits for either driving or riding, though rope or leather types of bridles may have been used earlier but such artifacts wouldn't survive. Evidence of the use of chariots dates to 2,100 BCE, in a region between Russia and Kazakhstan. Later evidence shows the use of chariots in regions in Greece, Egypt and China. The development of the chariot itself, as well as a specialized yoke for horses (different in the chest/neck area from an ox) supports their role in early transportation and work.

Horses have been used in warfare for over 4,000 years. Improvements in the chariot including the wheel and axle were made in the Bronze Age and the development of the bow facilitated the use of the chariot in warfare. The Hittites were known for their use of Chariots, and Kikkuli wrote a manual on conditioning the chariot horse. Cavalry warfare (horseback warfare) was popularized by the Scythians who mastered the art of turning backwards to shoot arrows over the horse's rump while riding. Alexander the Great and his horse Bucephalus were famous in leading the Greek cavalry. The Chinese also employed a cavalry to ward off the invading Huns. Horses were later used to carry cannons and other heavy artillery.

While the Romans relied mostly on an infantry (foot soldiers) for their warfare, the horse was extremely important for communication, and as such, extensive road systems were developed. The Romans also popularized horse racing, and even had gambling. Hipposandals, or the first horseshoes, were developed by the Romans to facilitate travel on the roads. During this time, horses were being bred for different purposes, and breeds were being established.

Into the Middle Ages (5th – 14th centuries), horses were used for war, agriculture, transportation and sport. During this time, horses were also bred specifically for their respective purposes: War horses (destriers, chargers, coursers and rounceys), draught horses, cart horses and riding horses (palfreys). The role of the horse continued to develop for agriculture and transportation through the Renaissance period (14th – 17th centuries) though with a greater interest in the science and art of the horse.

© Dennis Hallinan / Alamy Stock Photo

Section 2.3 Horses Today

In North America, the horse went extinct approximately 8,000-12,000 years ago, likely due to being hunted. The presence of horses here today is thanks to the Spanish and Colonial invasions. Thousands of horses were brought to North America over the years. Some horses escaped and developed feral populations in several regions (including the Mustang horses in the west and several island populations such as the Horses of Shackleford Banks in North Carolina). Horses were also traded with Native Americans who developed the horse for their own uses.

© Tom Plesnik/Shutterstock.com

Horses were used in America for transportation (both commercial and personal), exploration, farming and ranch work, mining, fire protection, war and policing, and for pleasure and sport. The importance of agriculture in the development of America is supported by the establishment of the Morrill Land Grant of 1862, which lead to the opening of State agricultural colleges, or "Land Grant Institutions". North Carolina State University is an example of a Land Grant Institution. The use of horses in pleasure and sport has led to roles for the horse in the Olympics, and the popularity of rodeo events and racing.

© Michael Klenetsky, 2011. Used under license from Shutterstock, Inc.

© Terry W Ryder, 2011. Used under license from Shutterstock, Inc.

According to the Animal & Plant Health Inspection Service, there were 3.6 million horses in the USA in 2012.

In North Carolina, the equine industry has a total economic impact of 1.9 billion dollars with more than 300,000 horses.

CHAPTER 3

Classification of Horses

The objectives of this section are to learn about how to describe horses based on type, breed, color, age and weight. This is important because it could be useful information when looking to purchase a suitable horse or to help identify a horse.

Section 3.1 Age, Height and Weight

Age

Knowing the age of a horse is important for determining its suitability to ride, its nutrient requirements (particularly for growing horses, or for aged horses), and to give you an idea about how many years it may have left to live. Many horses live well into their 30's with good veterinary care and nutrition.

Knowing how old a horse is depends largely on its documentation and paperwork, though this may not be available for many horses. Some breeds and disciplines require strict documentation such as a passport or registration papers that will identify the horse's birthday. Thoroughbreds are tattooed on their upper lip with their registration number, along with a letter of the alphabet that corresponds to the year they were born. In 2023 the list started back to the letter A.

A horse's age not only gives an indication of their maturity, but may also give an indication of their temperament and training. For example, a young horse is not only immature, but also likely has had less training. The term "broke" may be used to describe an older horse but more accurately describes one that has had extensive training. A "green" horse is one that is usually younger and has had less training. However, it should be noted that an older horse with little training may also be considered "green". Some competitions have classes especially designated for "green" horses based on their experiences.

It is also possible to get an estimate of a horse's age by looking at a horse's teeth. A horse is born with its teeth embedded within its skull, and as the horse ages the teeth erupt through the gumline. With wear (grinding, usually from feeding) the teeth are continuously worn down as they emerge. Depending on the extent of wear on the teeth, the structures of the visible tooth, and the presence or absence of mature or baby teeth, one can estimate a horse's age.

Terminology

Deciduous teeth – baby teeth

Permanent teeth – mature teeth

Incisors – the teeth at the front and center of the mouth used mostly for tearing grass. Labeled as corner, middle or intermediate and central going inward from each side.

Wolf teeth – the first set of premolars that are often removed to prevent interference from the bit, usually found only on the upper jaw, however horses could have upper and lower wolf teeth.

Canines (tusks or tushes) – teeth that are normally only found in male horses and are found behind the incisors, on both the upper and lower jaw. Some mares will have 2 canines, though it is rare.

Molars (and premolars) – cheek and jaw teeth that are used primarily for grinding

Wolf teeth – the first premolars, these are often removed shortly after they erupt so they don't interfere with the bit. They are usually only found on the upper jaw.

Bars – the interdental space between the incisors and the premolars

Infundibulum or Cup – a hollow rectangular or oval shaped depression on the surface (table) of the permanent incisors. As the horse ages, the infundibulum is worn and disappears

Dental Star – the pulp cavity or pulp mark that is exposed with significant wear (age). They first appear as a dark line in front of the cup and later become more oval in shape.

Galvayne's Groove – a groove that appears as a vertical line in the corner incisors that first appears at approximately 10 years of age

Number of Teeth

Most young horses have 24 "baby" teeth, which will fall out as they age (just like our own do). Most adult horses have between 36 and 42 adult teeth, with females tending to have fewer (because they don't usually have canines), and depending on if the wolf teeth are present or not.

Temporary (baby/ deciduous) teeth

$I - 3/3\ C - 0/0\ P - 3/3\ M - 0/0 = 12 \times 2 = 24$

Adult (permanent teeth) − MALE WITH 4 CANINES but Wolf Teeth removed.

$I - 3/3\ C - 1/1\ P - 3/3\ M - 3/3 = 20 \times 2 = 40$

I–incisor; C–canine; P–premolar; M–molar

Upper jaw/lower jaw × 2 for each side of mouth

*Mares without wolf teeth or canines would have 36 teeth.

The teeth can be used for aging because changes occur over time, including the presence or absence of mature or deciduous teeth, changes to the occlusal (grinding) surface (including the presence or absence of the cup or dental star), the angle of incisors, the shape of the teeth and the presence or absence of features such as Galvayne's groove.

Most horses are born with their first (central) incisors, and the rest of the incisors continue to erupt within the first few weeks to months. A horse begins to lose its baby teeth, which are replaced by their permanent mature teeth at about 2 and a half years of age. By the time a horse is five they are said to have a **full mouth** in wear, meaning all of their mature teeth are present and are long enough to grind and wear.

Teeth	Eruption	Teeth	Eruption
Deciduous (Baby Teeth)		*Permanent Teeth*	
1st incisor	Birth–1 week	1st incisor	2½ years
2nd incisor	4–6 weeks	2nd incisor	3½ years
3 incisor	6–9 months	3rd incisor	4½ years
2nd premolar	Birth–2 weeks	Canine	4–5 years
3rd premolar	Birth–2 weeks	1st premolar (wolf tooth)	5–6 months
4th premolar	Birth–2 weeks	2nd premolar	2½ years
		3rd premolar	3 years
		4th premolar	4 years
		1st molar	9–12 months
		2nd molar	2 years
		3rd molar	3½–4 years

Appearance of deciduous and permanent teeth.

Equine teeth are not the same the entire length down the shaft. Therefore, depending on what part of the tooth is visible above the gumline, there may be identifiable features in the dental structure based on what is typical for a horse of a given age.

At the grinding surface, the infundibulum or cup is initially visible in the mature teeth. This is a little cavity in the tooth, similar to, though smaller than, what you might feel on your molars. The cup is also often filled with old feed and may appear to be a different color. As those teeth are worn, the cup disappears. The cup disappears from the lower central, intermediate and corner incisors at 6, 7 and 8 years, and from the upper central, intermediate and corner incisors at 9, 10 and 11. Thus, at age 11, a horse is often said to have a "**smooth mouth**" as all of its cups are gone.

The dental star or pulp cavity is starts to appear around the age of 6-8 at the lower central incisors and is present on all incisors by approximately 12 years of age. It gets bigger and appears darker as the horse ages.

When both the cup and dental star are present (between 6-11 years of age) – the dental star will be closer to the outside (lip side) of the tooth, with the cup closer to the tongue side.

The angle of the incisors when looking at them from the side increases (the teeth become more slanted, and less vertical) with age.

The shape of the teeth also changes with age. In younger horses they are oval (sideways to the horse's mouth; transversely), then become round in shape by the time a horse is 10, then become triangular in shape around 16 years of age and finally become oval again (front to back; rostro-caudally) by 20 years of age.

Some horses also develop a hook on the upper corner incisor around the age of 7. They may also develop another hook at 13 years of age.

Galvayne's Groove, a dark vertical line appearing in the upper corner incisor begins to appear at age 10. By the time a horse is 15 it is halfway down the tooth, and extends all the way down the tooth by 20 years of age. At 25, it is present only on the bottom half of the tooth and is gone completely by 30 years of age.

Height

A horse's height is measured in hands, where 1 hand equals four inches. Height is measured from the ground (without horse shoes) to the horse's withers. See photo below.

> Example: 13.2 hh (hands high)
>
> = 13 hands plus 2 inches
>
> = 54 inches (13 $\times$ 4 + 2)
>
> (one would never write 13.5 hh)

> An average horse is about 16 hh.

Weight

Body weight can be most accurately determined using specialized equine scales, though these may be expensive for personal use. Estimates can therefore be made with the use of a weight tape that is wrapped around the horse's heart girth area and has indications for weight estimates when the ends are brought together.

Alternatively, one can use a calculation using the horse's heart girth and body length. This equation is more accurate for estimating a horse's length than a weight tape because the horse's body length is also considered.

$$\text{Kg BW:} \quad \frac{\text{Heart girth (cm)}^2 \times \text{length (cm)}}{11,880}$$

$$\text{Lbs BW:} \quad \frac{\text{HG (in)}^2 \times \text{length (in)}}{330}$$

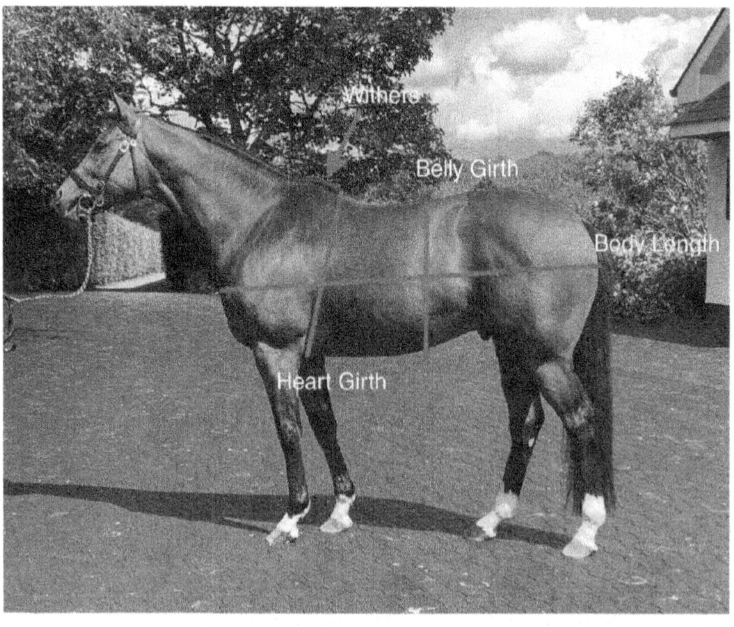

In addition to determining how much a horse weighs, information regarding how much fat coverage has may be important as well, particularly in relation to the horse's overall health. Taking body measurements over time (such as heart girth or belly girth circumferences, see photo) can indicate if a horse is losing or gaining weight as fat.

There are many different ways to assess fat coverage including using an ultrasound, but the Henneke Body Condition Score is an easy and common scale to estimate fat coverage. This is a 1-9 scale where a 1 is a fully emaciated horse and a 9 is a grossly obese animal. The score is assigned based on examination and palpation of regions where fat is often deposited, including the ribs, shoulder, neck, tailhead and withers.

Ideally a horse is between a score of 5-6, but it truly depends on their use and overall health. Broodmares tend to do better with a little extra condition, averaging closer to 6-7, while athletes tend to be lower in the 4-6 range (also depending on discipline). A horse is considered overweight if they are greater than a score of 6, and obese if they are greater than a 7. Similarly a horse would be considered underweight if they are less than a 4.

Body condition score of 3.5 (notice visible ribs and outline of pelvis and withers)

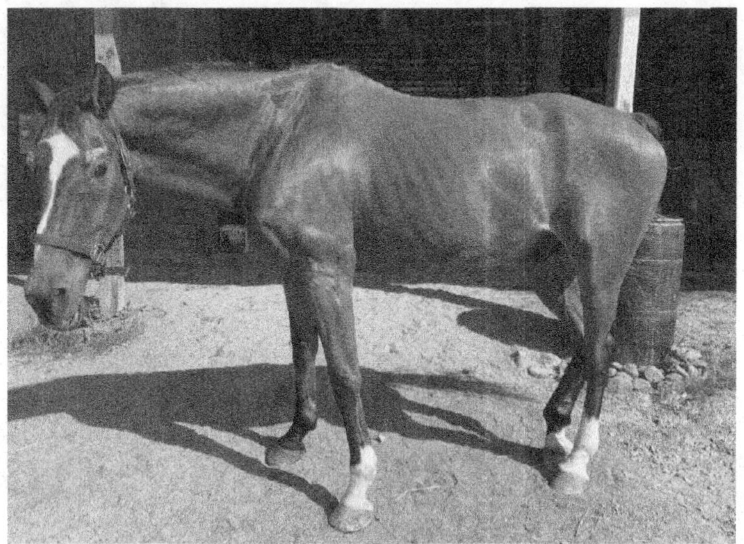

Photos courtesy of author

Body Condition Score of 4.5 (notice faint outline of ribs)

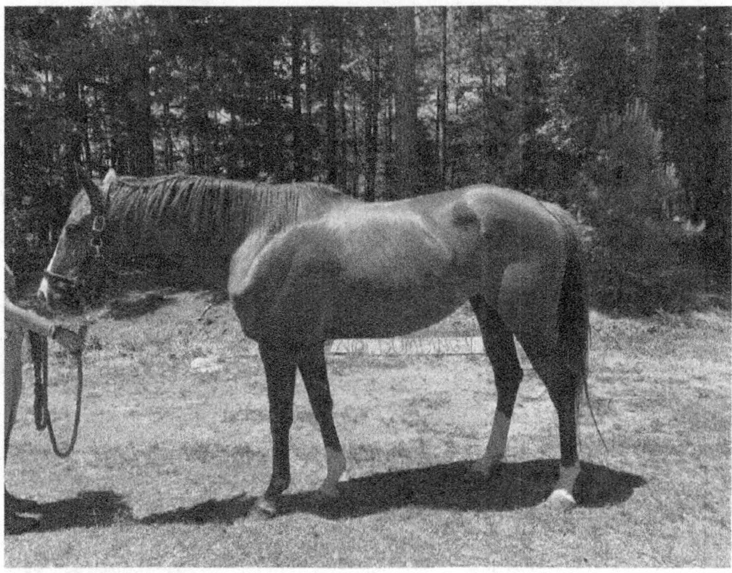

Body condition score of 5

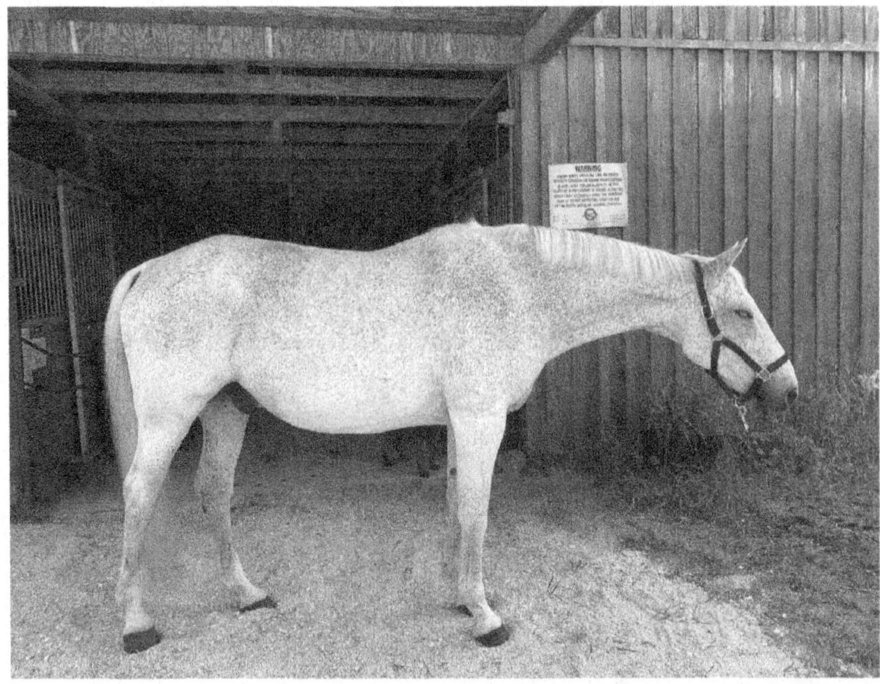

Body Condition score of 6

Body condition score of 7 (horse) (notice thick neck, fat pads behind shoulder, rounded buttock)

Body Condition Score 8 (pony) (notice crest along neck, fat pads on hindquarters)

Section 3.2 Equine Colors

Based on genetics, there are two main colors of horse hair; black and red (chestnut or brown), as well as lack of color (white). Then, there are color modifiers that can affect the location of the colors of red or black hair; can dilute or lighten these colors, or can whiten the hairs (both progressive whitening as seen in gray horses, or localized white patterns).

Any color of horse may also have white markings described below.

> **Black** – All of the hair is black. Occasionally it may have a brown or red glow to it due to sun bleaching.
>
> **Bay** – The body color varies from light to dark brown, but has black points such as the mane and tail and often the lower legs. Variations of bay include light bay (very light brown body), dark bay (which may even resemble black) or dapples (spots of lighter and darker circles).
>
> The Thoroughbred industry may describe a horse as "brown", which generally describes a dark bay. There is a rare genetic variation that results in a brown horse – one that is not technically bay but also not black.
>
> **Chestnut** – Often called sorrel in some disciplines. The horse has a brown body, mane and tail. The brown color may vary extensively, ranging from red or golden brown, to more of a dark liver brown, called a liver chestnut. Other variations include the color of the mane, which may be lighter and flaxen in color or darker (but not black).
>
> **Gray** – A horse with a mixture of white hairs interspersed with another (usually dark) color. The horse is usually born with a solid (non gray) coat and will "gray out" with age, eventually becoming white. Gray horses have black skin. Many gray horses will have dapples, and the age at which a horse grays varies greatly, with some horses becoming all white at a young age, while others stay more of a steel gray color later in life.
>
> A variation of gray is the Silver Dapple gray, which is a unique gray caused by the silver dapple gene, and is not considered a typical gray.

White – A white horse has white hair color and pink skin. They may have patches of colored hair. Genetically, there are two main ways a horse is truly white; through the sabino gene which is a pinto pattern that is expressed all over the horse's body, or a dominant white gene which is relatively rare. Contrary to popular belief, there are no albino horses. Some gray horses, or horses with various forms of dilutions (see below) are often mistaken for white horses.

Dilutions are the result of the dilution genes (units of information from each parent); if a horse has one dilution gene (ie. They inherited the trait from one parent) they will have a single dilution and if they have two dilution genes (ie. They inherited the trait from both parents) they will have a double dilution.

Base Color	Single Dilution	Double Dilution
Black	Smoky black	Smoky cream
Bay	Buckskin	Perlino
Chestnut	Palomino	Cremello

Buckskin – A yellow, golden or tan body color with dark points (mane and tail)

Perlino – A pale cream or light tan color body, slightly darker mane and tail (with a reddish tinge) and having blue eyes.

Smoky black – a light or smoky black color body

Smoky cream – very similar to perlino and often only distinguished by genetic testing

Palomino – golden or tan coat with blond mane and tail

Cremello – A pale cream body color with mane and tail often the same color, often with blue eyes. Commonly mistaken for white.

Roan – White hairs are evenly mixed with the horse's base color. Typically the heads are solid colored and have less white hairs. Roans do not change color throughout their lives (unlike grays). Further descriptions can be made based on the coat color:

> **Red (Strawberry) Roan** – Chestnut body color

> **Blue Roan** – Black/brown body color

Dun – Caused by the dun gene, a horse with a body color that is gold or yellow and a mane and tail that depends on the base color. Duns typically have a dorsal stripe and often have primitive markings on their legs and a transverse stripe at their withers. Variations include Grullo (grulla or blue dun) where the horse is black and the coat color appears mouse-colored, red dun where the horse is chestnut resulting in a tan color body with red mane, tail and markings, or bay dun, where the body is yellow or tan with black mane, tail, legs and markings.

Appaloosa colors refer to a mix of colors caused by the leopard gene complex that is found predominantly in the Appaloosa breed. Such variations include blanket patterns, spotting leopard patterns, frost or snowflake patterns.

http://www.appaloosa.com/registration/indentify.htm

A pinto refers to the color of a horse when their body has both larger white areas and any other color (usually black/bay or brown/chestnut).

In the UK, more specific terms such as skewbald and piebald are used to describe a horse with brown and white or black and white colors respectively.

Pinto colors depending on genetics, and are described based on the pattern of the white and the other color – these patterns include overo, sabino, tobiano or tovero.

Overo – a mix of any color and white where the color tends to be OVER the horse's back. They usually have dark legs and the tail is usually one color. The color pattern tends to be irregular/blotchy.

Tobiano – a mix of any color and white but the color is more regular with crisp lines, compared to overos. The color is usually on the belly or flank area and the legs are usually white. The tail is often a mix of two colors.

Tovero – a mix of tobiano and overo characteristics due to a horse having the genetics for both overo and tobiano patterns.

Sabino – a spotting pattern that often has white high on the legs, extensive white markings of the face and belly splashes. A maximal sabino has white over most of the body resulting in an almost or even all white horse.

Markings

Concentrated areas of white may also appear on a horses face or legs, and can be used to identify horses. For example, you might describe a horse as a chestnut with a blaze and two hind socks.

Head Markings

Blaze – a wide white stripe down the center of the horse's face, wider than 1 inch

Stripe or strip – a narrow white stripe down the center of the horse's face

Star – a white marking between the horse's eyes

Snip – a white marking between the horse's nostrils

White face – when the white covers the entire face including the eyes

Leg Markings

White appearing on the legs is generally named based on the region of the horse's legs where the white ends.

Coronet – white around the horse's coronet region

Pastern – white up to the pastern region

Fetlock – white up to the fetlock area

Sock – white that extends into the cannon area above the fetlock but below the hock or knee

Stocking – white that extends to the knee or hock

Other terms may be used to further describe markings such as "partial", "interrupted", "irregular", etc.

Other markings include bend-or spots which are dark spots, ermine marks (dark spots on white markings), brindle patterns, and belly spots or splashes.

3.3 Types of Horses

Horses can be classified based on body size, weight and build, or based on temperament. If based on size, we use the term "hand" (4 inches) to describe heights. We also categorize horses as light horses, draft horses and ponies when describing basic sizes. Alternatively, we can categorize horses based on temperament such as hot-blooded, warmblooded or coldblooded animals.

Light horses are typically 14.2 – 17.2 hh (hands high) and weigh between 900-1,600 pounds. Examples of light horse breeds include Quarter Horses, Thoroughbreds and Hanovarians. Draft horses may be even taller, but more importantly weigh upwards of 2,000+ lbs. Examples of draft horses include the breeds Belgians, Clydesdales and Percherons. Ponies are less than 14.2 hh and usually weigh between 500-900 lbs. Examples of pony breeds include Shetlands and Welsh ponies.

When categorizing based on temperament, the terms hot blood, warmblood and cold blood do not refer to their actual blood temperature. Rather, a hot blooded horse tends to be more spirited and active, while a cold blood horse tends to be more docile. Examples of hot bloods include Arabians and Thoroughbreds, Warmbloods tend to refer to European breeds such as the Hanoverian, Holsteiner and Dutch Warmblood. Coldblooded horses are typically draft breeds and ponies.

Donkeys are a common equid species that are also known as asses. A Jack is an adult male donkey while a Jenny (or Jennet) is an adult female donkey. Donkeys don't usually have chestnuts and have distinctive voices ("hee-haw"). Mules and hinnys are crosses between horses and donkeys. A mule is a cross between a mare and a Jack while a hinny is a cross between a stallion and a donkey.

Feral populations of horses are descendants of horses that were domesticated at some point, but are now considered wild. There are habits of wild horses in California, Colorado, Idaho, Montana, Nevada, New Mexico, Utah, Wyoming and North Carolina (and others). Many of the populations of horses on the Outer Bank islands of North Carolina are descendants of horses that came across the Atlantic from Spain in boats that had crashed and the horses swam to shore.

Miniature horses are those that stand 34 inches and less. They are not considered ponies, but are rather a scaled down version of a horse. Miniature horses are used for driving, as pets, and more recently are popular as guide animals. http://www.guidehorse.com/

© Lilac Mountain, 2011. Used under license from Shutterstock, Inc.

© Lilac Mountain, 2011. Used under license from Shutterstock, Inc.

Section 3.4 Breeds

A breed is defined as a group of horses selected for their common ancestry and common characteristics. Breeds tend to "breed true" and pass on desired genetic traits to their offspring. Purebreds are horses with distinct bloodlines within a breed, while some breed registries have open studbooks and allow any horse to register provided they meet specific criteria. The common characteristics of a breed include conformation such as body structure, athleticism, or color specifications. Color registries are those where a horse can be any true breed and have any conformation type or ancestry, but must have a certain color, such as those registered as Palomino.

Some characteristics of a few breeds are listed, though additional information about different breeds can be found:

http://www.ansi.okstate.edu/breeds/horses/

Andalusian. The Andalusian is a Purebred Spanish Horse that tends to be gray in color, though some are bay or black, and typically range 15-16.2 hh. This breed is a major genetic influence on the Lipizzaner breed.

Appaloosa horses are usually 14 – 16hh, 950 – 1300lbs and were bred by the Nez Perce Indians. They have three key characteristics; white sclera of the eyes that are visible, mottle skin at the muzzle and genitalia, and stripped hooves. They also have unique coat color patterns including the blanket, blanket with spots, roan, leopard, etc.

Arabian horses were developed in Arabia by Bedouins. They were bred for endurance and are often still very successful in endurance competitions. They tend to be smaller, at 14.1 – 15.2 hh. They typically have a dished face, large prominent eye, small teacup muzzle and they carry their tails high. They are considered the oldest purebred horse.

Belgian horses are a draft breed of horse that tend to be chestnut in color with a blond mane and tail and they originated in Belgium. They can reach upwards of 17hh and 2000 lbs. They may be afflicted with a genetic disorder called junctional epidermolysis bullosa.

Clydesdales are a draft breed named for a district in Scotland where they were established at the River Clyde. They are usually 16 – 18hh, 1800-2000 lbs, and are often bay with white face and legs, and may have sabino markings such as belly splashes. Clydesdales are most commonly recognized as the "Budweiser Horses".

Dutch Warmbloods belong to the Koninklijk Warmbloed Paardenstamboek Nederlan (KWPN) registry that governs their breeding. They have a rigorous selection process that involves Keurings or inspections. They tend to be excellent sport horses for dressage or show jumping and average about 16hh.

Friesian horses are from the Northern Netherlands and usually stand 14.2 – 17hh. They developed from Medieval war horses and have what is known as a Baroque body type. They are only black in color and tend to have long flowing manes.

Haflingers originated from Hafling, Austria, which is now part of Italy. They are small at 13.2 – 15 hh and are always chestnut in color with a white mane and tail. The first horse to be cloned was a Haflinger.

Hanoverians are a German warmblood breed from Hanover Germany. They have rigorous breed qualifications and are known as excellent sport horses. They average 16-17hh.

Icelandic horses are from Iceland and tend to be 13 – 14hh and 700 – 850 lbs. They have unique gaits called the Tolt which is a running walk, and the Skold which is a flying pace.

Lipizzans (also called Lipizzaners) originated in Lipizza, from Spanish horses such as the Andalusian. The Spanish Riding School was established in 1735 and is currently in Vienna, Austria. They are gray (white) in color, but are born dark. They are known for some unique movements of classical dressage called the airs above ground.

Morgans are an American breed named for a horse called Justin Morgan who lived in the late 1700's, whose name was actually "Figure" at birth and Justin Morgan was the owner. Morgans tend to be chestnut, bay or brown. They are major breed influencing racing Standardbreds and Tennessee Walking horses. They are used as either harness or riding horses.

Norwegian Fjords or just "Fjords" are ponies (typically 13-14hh) from Norway that have primitive coloring, including a dun body color, with a black mane on the inside with white on the outside.

Paints or the American Paint Horse refers to a breed of horse that most often has pinto coloring (white plus any other color spotting pattern). These horses have a stocky body type and to be registered with the American Paint Horse Association both the sire and dam must be registered with the American Paint Horse Association, The American Quarter Horse Association or the Jockey Club (Thoroughbred). They must also meet color requirements.

(Note: This differs from a pinto, which simply refers to the color pattern, though there is a color breed registry for pintos of any breed through the Pinto Horse Association)

Paso Finos are a breed from Puerto Rico and Columbia, and are similar to the Peruvian Paso from Peru. They have a unique natural gait called the Paso.

Percheron are from the La Perche region of Normandy France. They are a draft breed that tend to be black or gray in color. They differ from some other draft breeds because they do not have excessive feathering (long hairs) at their fetlocks.

Quarter Horses (The American Quarter Horse) was developed to run a quarter of a mile. They are also popular for rodeo events, ranch work and other Western riding disciplines. They are highly versatile in their uses. They are the first breed native to the U.S. They have a stock-type body with muscular conformation, compact body type, short cannon bones and hocks that are low-set to the ground. They tend to be 15.2 – 16hh.

© Ron Rowan Photography, 2011. Used under license from Shutterstock, Inc.

Saddlebreds or the American Saddle Horse was developed for easy riding. The foundation sire (to which all Saddlebreds can be traced to) was named Denmark. Some perform a unique slow gait (with four beats) or a rack.

Shetland Ponies are a small pony breed (<46in or 11.2h). They were often used in mines due to their small body size and hardiness. They are very popular with children.

Shires are a tallest of the draft horses that originated in England (Lincolnshire, Yorkshire…). They tend to be black, gray, chestnut or bay and weigh up to 2,200 lbs.

Standardbreds are the American Trotting Horse and are named for trotting a mile in less than "standard" time. There are two foundation Sires: Messenger (who was a Thoroughbred) and Hambletonian 10. They are used in harness racing, at either the trot or the pace.

Tennessee Walking Horses originated in Tennessee (registry began in 1935). They were bred from Standardbreds, Morgans, Thoroughbreds, pacers. The foundation sire is said to be Allan F-1 (aka Black Allan). The Grand Champion Walking Horse is awarded in Shelbyville Tennessee each year. Tennessee walking horses may be flat-footed pleasure horses, or performance horses that are shod with elevated pads (up to a maximum of 50% of their natural shoe length). They have unique gaits that include the running walk and the rocking horse canter.

Thoroughbreds can be traced back to three foundation sires, Darley Arabian, Godolphin Arabian and Byerly Turk, who all foaled between 1679 – 1724. The general stud book was established in 1791. Thoroughbreds are most recognized as racing horses with typical distances of 6 furlongs to 1.5 miles. They tend to be 15-17hh.

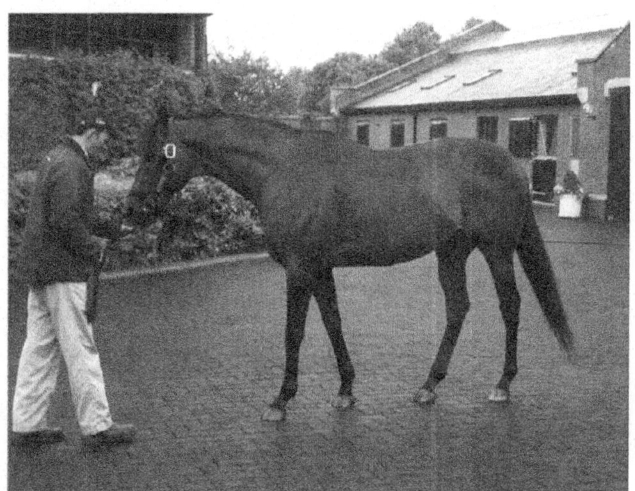

Photos courtesy of author

Welsh Ponies are from Wales and are slightly larger than Shetlands (13-14hh) and are popular riding ponies for kids.

There are also several rare (fewer than 5,000 horses worldwide) and critical (fewer than 2,000 horses worldwide) breeds.

Rare

- Canadian
- Lipizzaner
- Rocky Mountain

Critical

- Akhal Teke
- Spanish Mustang
- Suffolk
- Cleveland Bay

CHAPTER 4

Form and Function

The objectives of this section are to learn about how the unique features of the horse's body work together to allow him to be athletic (or not!)

Section 4.1 – Conformation

Conformation describes the way the horses' body is put together. Conformation is important because it influences the usefulness of a horse, as a horse with poor conformation will be more likely to develop injuries and lameness.

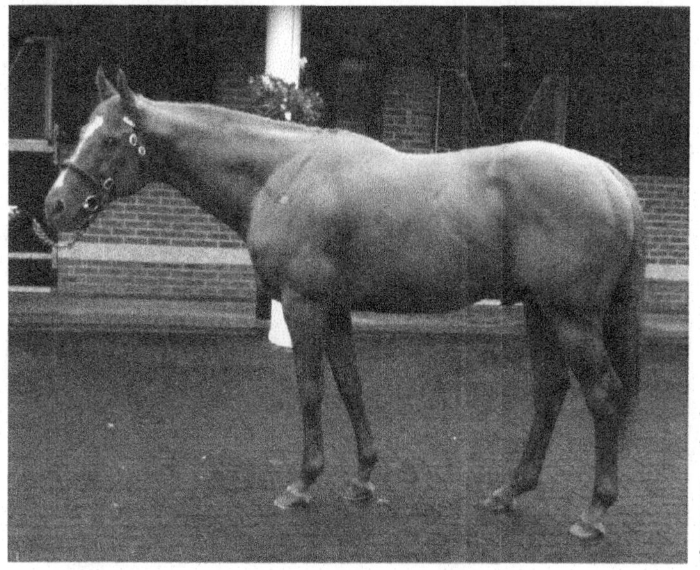

A thoroughbred with good conformation

The conformation assessment takes into consideration the profile, front and back views and movement, type, structural correctness, muscling and size. The body should be balanced and topline and underline considered. Type refers to the type or breed of horse being examined; if it is an Arabian, it should have typical features such as a dished face and large eyes, while a horse that is to be used as a hunter should have a long and slender neck, and a stock horse should have a short back and be well muscled.

The head should be in proportion with the rest of the body. It should not have any curvature to the face. A forehead that is concave (dished) is common in Arabians, while a forehead that is convex (Roman nose) is common in Shires; but neither are desired in many other breeds. The eyes should be alert and large, set well apart on the head. The poll and jaw should have ample room for flexion, as there should be room for two fingers between the jaw bone and the wing of the atlas (top bone in the neck). The teeth should align and there should be no monkey jaw (bottom jaw extends beyond the upper jaw) or parrot mouth (upper jaw extends beyond the lower jaw). Minor tooth defects may be corrected surgically, but a very misaligned dental structure will prevent a horse from eating properly, and can lead to a life of poor condition and euthanasia may be considered.

The neck should be in proportion to the head and body, and well-set into the withers and shoulder. There should be two equal curves of the vertebra of the neck, and it should appear to enter the skull from the top. If a horse has a large head, it should have a shorter, stronger neck to support it.

The chest should be broad to allow for room for the heart and lungs. The withers should be high and well-defined to allow for a good fit for the saddle. It is common for older horses to develop a bit of a sway (sunken) back making the withers appear more pronounced. This is due to gravity over time stretching out the ligaments along the horse's spine, as well as loss of muscle tone.

The shoulder should be approximately a 45 degree angle to the horizontal. More sloping shoulders tend to result in a longer stride, while higher upright shoulders lend to more leg action. Therefore, an ideal shoulder may depend a little on the discipline.

The forelimb should be viewed from the front and side. From the side, a line should be able to be drawn from the shoulder, through the mid-elbow, through the knee and fetlock. The hoof should hit the ground between the point of the shoulder and the point of the elbow. From the front, the line should drop from the point of the shoulder and bisect all limb bones and hoof.

Forelimb Faults

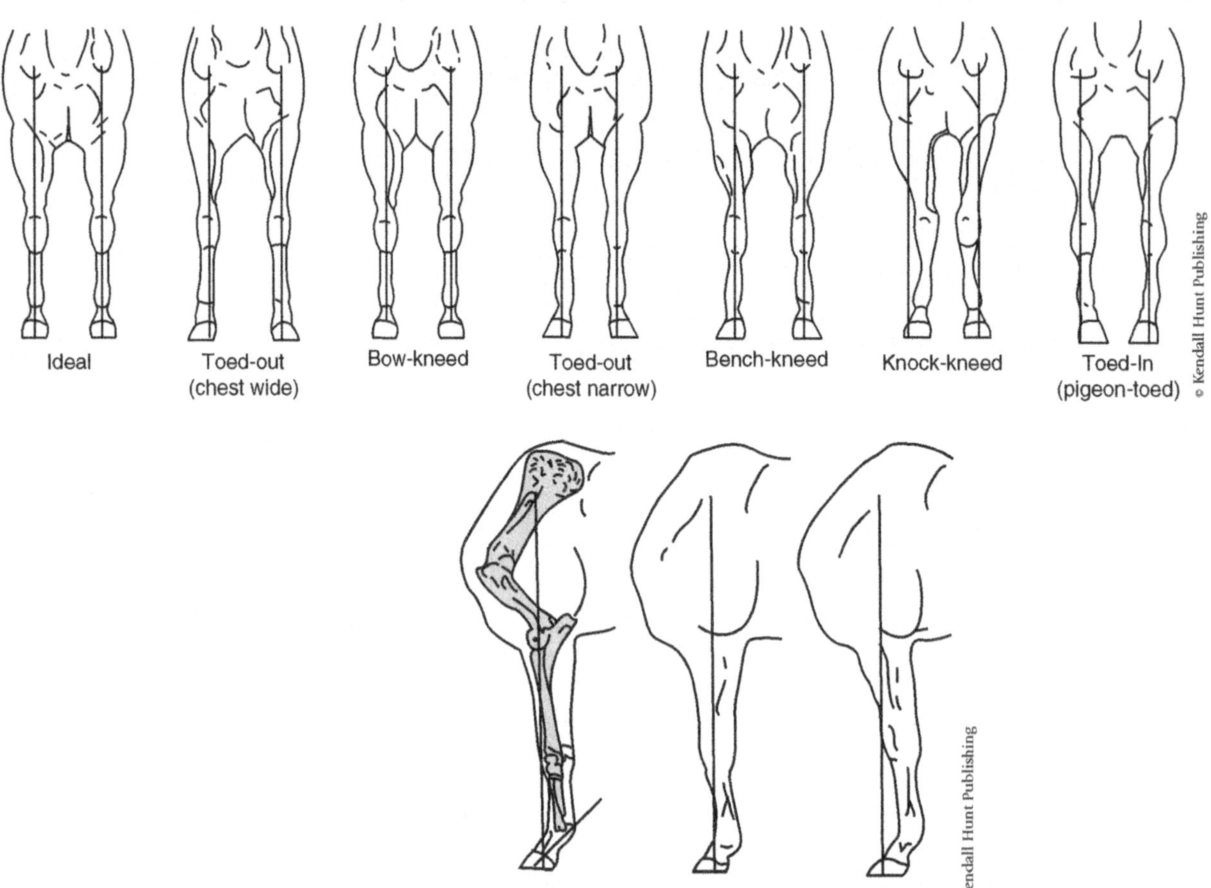

Bench Knees – cannons fail to come from the centre of the knees (front view)

Bow Knees – Horses stand over the outside of their front feet, a bowed out appearance. Can cause side bones (front view)

Knock-knees – horse stands in at the knees or is too close at the knees (front view)

Pigeon-toed – toed in (front view)

Splay footed – toed out (front view)

Buck knees/Knee Sprung – over at the knees (side view)

Calf knees – knees that break backwards (side view)

Camped out/under – the entire leg is pushed out or under the vertical line from the shoulder (side view)

In the feet, the angle of pastern should be close to angle of shoulder. In addition, the hoof angle should be the same as the pastern.

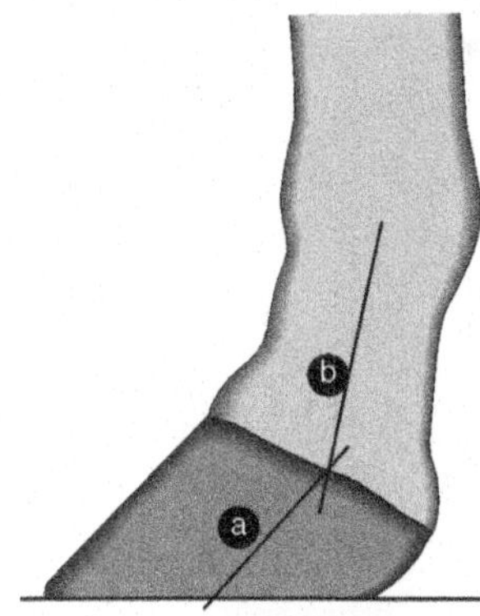

A. Broken foot in which the foot angle (a) is less upright than the pastern angle (b)

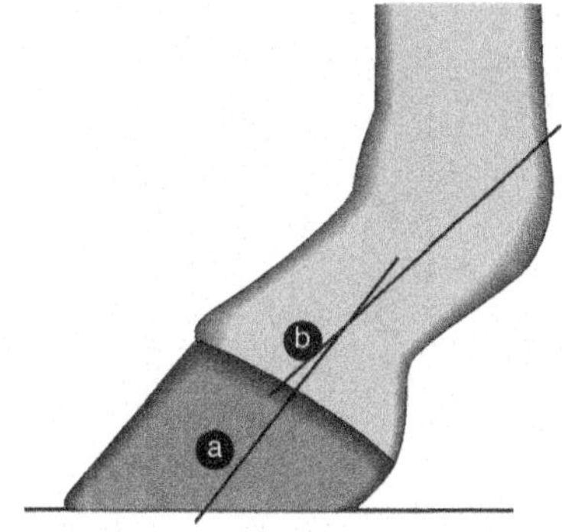

B. Broken foot in which the foot angle (a) is more upright than the pastern angle (b) ("coon foot")

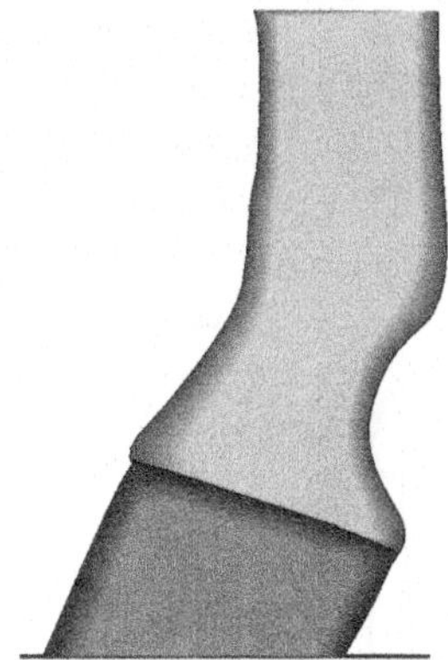

C. Club foot (foot angle is 60° or more)

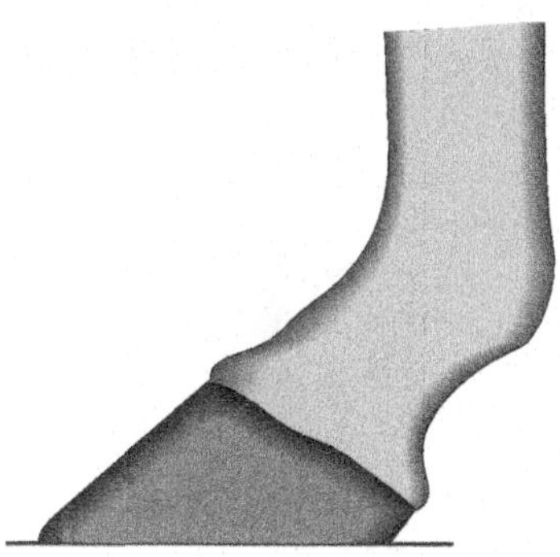

D. Sloping foot (foot angle is less than 45°)

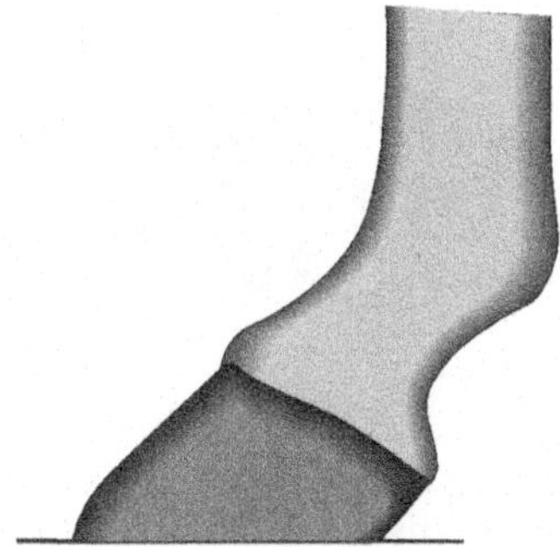

E. Bull nosed foot

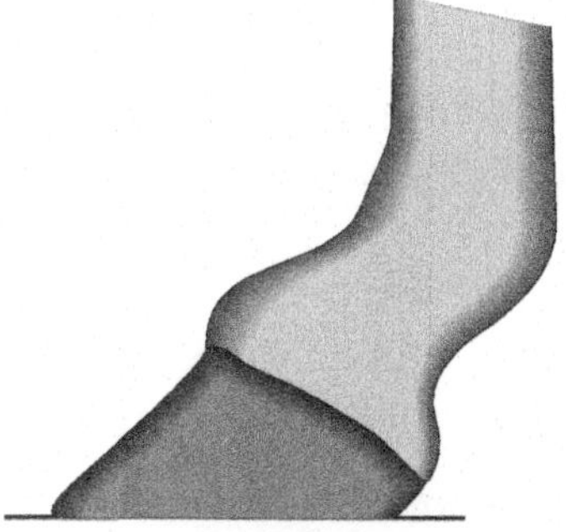

F. Buttress foot

Foot Faults

Broken foot – hoof angle less upright than pastern

Broken foot – hoof angle more upright than pastern (Coon foot)

Club foot – short upright hoof and pastern angles

Sloping foot – both hoof and pastern too sloped

The back should be in proportion with the body. If it is too long it may be weak, and too short may result in back pain. The hindquarters should have a rounded rump with good length from the point of the hip to the point of the buttock. Arabians tend to have a very flat rump.

In the hind legs, the hock should be inline with the point of the buttock and the fetlock and gaskin well muscled.

Vertical line from point of buttock should fall in center of hock, cannon, pastern, and foot.

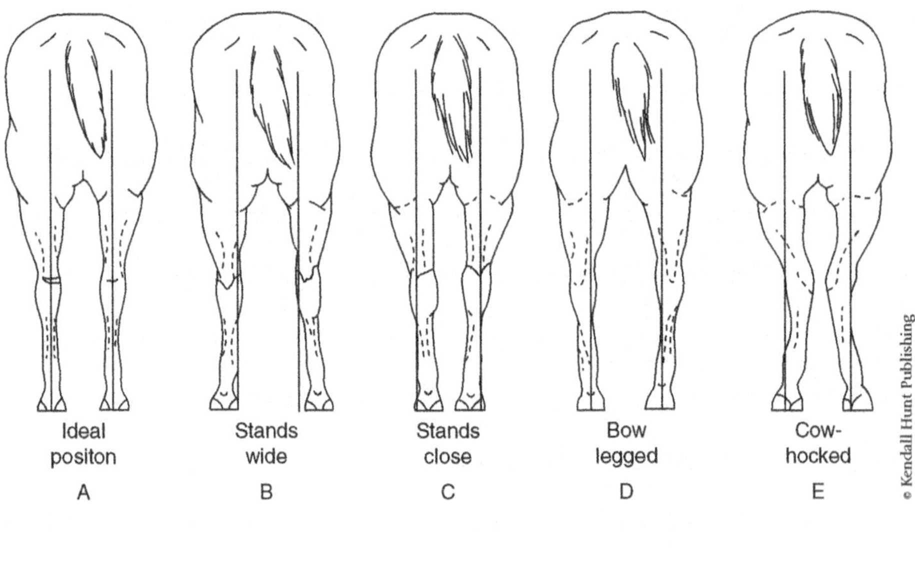

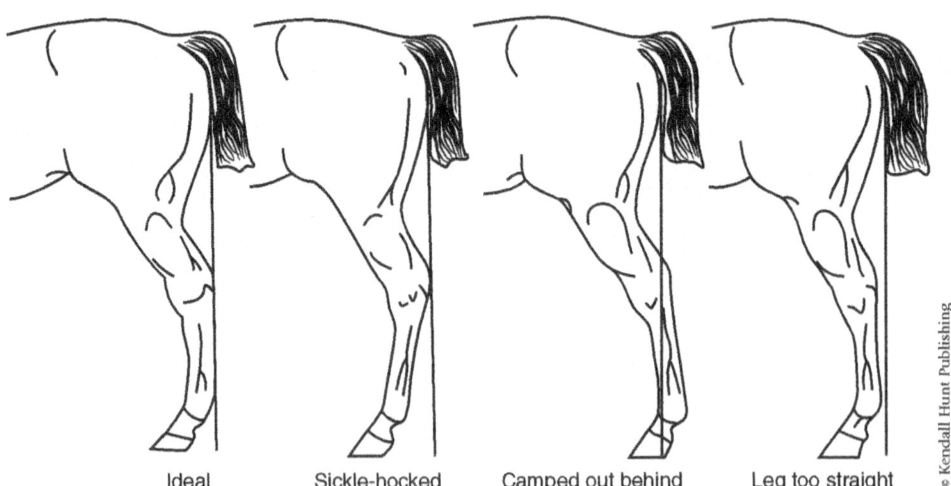

Hindquarter Faults

Bowlegged – when a horse stands pigeon-toed on his hind feet with hocks turned out (back view)

Cow Hocked – Horse stands with point of hocks turned inward, base wide and splay footed (back view)

Post-legged – too little angulation in the hock (side view)

Camped out – rear legs set out behind the hip (side view)

Camped under – rear legs set too far under (side view)

Sickle hocked – when horse's rear legs have too much angle at the hock and resemble a sickle (side view)

Section 4.2 Gaits and Movement

There are four natural gaits: walk, trot, canter and gallop. There are several other gaits unique to different breeds including the pace, running walk, rack, foxtrot, and others.

Gait = Footfall pattern of how the horse moves his legs to move forward

Terminology

Lateral Movement = legs on same side move together

Diagonal Movement = legs on opposite side move together

Beat = time when foot (or feet) strike the ground

Stride = distance between imprints of the same foot

Step = distance between imprints of the two front feet

Action = refers to the flexion of knees & hocks and how high the feet are lifted

There are four natural gaits: walk, trot, canter and gallop. There are several other gaits unique to different breeds including the pace, running walk, rack, foxtrot, and others.

The walk is a slow four beat gait: left hind, left front, right hind, right front. Running walks unique to some gaits are the same footfall pattern but slightly faster.

The trot is a two beat gait, and is called a jog in Western disciplines and tends to be a little slower. The footfall pattern is: left hind and right front together, suspension, right hind and left front together, suspension.

The canter is a 3 beat gait. The canter (and gallop) has a "leading leg" (eg. Right lead or left lead), which refers to the front leg on the inside on the way of going, which tends to have a more pronounced movement. The terminology is a little confusing, as the "leading leg" is actually the last leg to hit the ground in the cycle of the gait. In Western disciplines, the term "lope" is used instead of "canter".

A left lead canter has the footfall pattern: right hind, left hind and right front together, left front, suspension. A right lead canter has the footfall pattern: Left hind, right hind and left front together, right front, suspension. A horse can perform a "flying lead change" if they switch from left to right lead (or vice versa) mid stride.

© Abramova Kseniya, 2011. Used under license from Shutterstock, Inc.

The gallop is a four beat gait that is very fast. The footfall pattern of a right lead gallop: Left hind, right hind, left front, right front, suspension.

When a horse is backing up, it is a two-beat diagonal movement that is the same footfall pattern as the trot, but is slower. The "back" is not considered a true gait, because it is not moving the horse forward.

Other gaits

The pace is a lateral two-beat gait, where both left legs and both right legs move together. It is a lateral rolling gait meaning the horse leans side to side. Standardbred horses are known for racing at the pace.

The running walk of Tennessee walking horses is faster than an ordinary walk but with the same footfall pattern. The hind legs however tend to overstep the front, and gives the horse a gliding motion. The horse also moves his head up and down with each stride.

The rocking horse canter of Tennessee walking horses is also the same footfall pattern of a normal canter but has very high action in the legs, is slower and gives a rocking horse motion.

Many gaits are considered "ambling" gaits, and horses that are able to do these gaits are often called "gaited horses". These gaits are four beat gaits, that are typically at a speed between a walk and a canter. Some ambling gates are lateral (for example hind left, front left, hind right, front right), while others are diagonal (for example hind left, front right, hind right, front left). The beats in which the hooves hit the ground may also be isochronous, where there is an even beat between each hoof – 1, 2, 3, 4. Other gaits are non-isochronous where two hooves hit the ground closer together, as in, 1-2, 3-4.

The rack of American Saddlebreds and some other breeds is a lateral four beat gait. The tolt performed by Icelandics is the same footfall pattern as the rack.

The Fox Trot is performed by Missouri Fox trotters and is a four beat diagonal gait – but the front foot hits the ground first.

The paso is a lateral ambling gait that is named based on its speed: fino, corto and largo.

Movement Faults include those caused by lameness, uneven pace, where the horse is uncoordinated or crooked, or has short and stiff strides.

Specific faults include:

> **Forging** – The rear foot strikes the front foot on the same side of the horse's body (hind right hits front right). Also called overreaching
>
> **Cross-firing** – The ear foot strikes the front foot on the opposite side of the body (hind right hits front left). This is most commonly seen in pacers.
>
> **Scalping** – When the front foot hits the hind foot at the hoof on the same side of the body (right front hits right hind)

May be further called:
- Speedy cutting (if the pastern is hit)
- Shin hitting
- Hock hitting

> **Interference** – When one foreleg strikes the other foreleg

Maybe be further called:
- Ankle hitting
- Shin hitting (not to be confused with previous!)
- Knee hitting
- Forearm hitting

> **Paddling** – Horses legs swing out with each stride. Also called winging out.
>
> **Dishing** – Horses legs swing in with each stride. Also called winging in
>
> **Plaiting** – Horse appears to be walking on a tight-rope!

Different types of boots or wraps may be worn by horses to help protect their legs from being hit by another. Bell boots are worn to help protect hoofs/shoes from horses that may overreach (forge).

CHAPTER 5

Equine Uses

The objectives of this section are to learn about how the horse is used in pleasure riding, sport and work.

Section 5.1 Preparing your horse for use

A halter is a piece of equipment worn on the horse's head while leading, tying and grooming. It has a piece that goes over the horse's poll, side pieces and a section that goes around the nose and under the jaw, though there are several different types of halters. Typically there are rings for attachment underneath the chin, and on each side of the horse's cheeks. A lead shank (or rope) is used to lead a horse and attaches to the bottom ring of the halter. Typically we lead horses by their left side, staying near the shoulder. Horses should only be tied to trusted objects, and cross-ties facilitate working around a horse. Cross ties are attached to the side pieces of a halter.

© Margo Harrison, 2011. Used under license from Shutterstock, Inc.

Grooming the horse not only prepares it for tack placement and being ridden but also can help with circulation, relieving muscle tightness through a massage, and can help identify injuries through regular inspection. Several tools are useful for grooming.

A curry comb is usually a hard rubber or plastic comb that is used in a circular motion to lift dirt and increase circulation and massage the horse.

A dandy brush is a hard bristled brush that is used to help remove dirt. A flicking motion can help lift dirt up and away.

A soft or body brush is used to help remove dust and smaller particles. It can be used on the face and legs.

A mane and tail comb or brush is used to remove debris from the mane and tail. A mane pull can also be used to thin and shorten the mane.

A hoof pick is used to remove dirt and other debris (rocks or pebbles) from the hooves.

Several other tools may be used as part of a grooming regime including a shedding blade, scissors, towels and sponges.

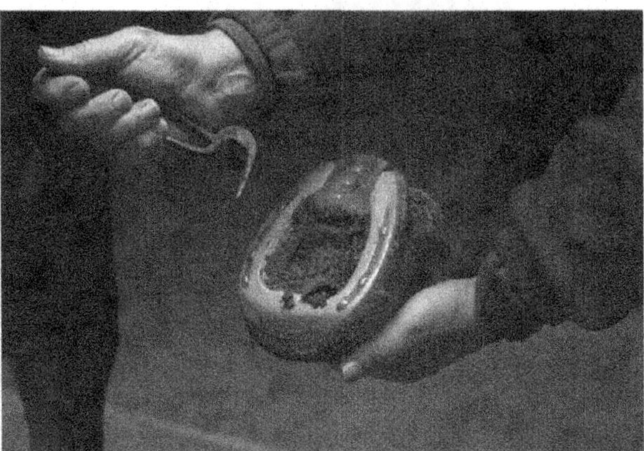

© Jeff Banke, 2011. Used under license from Shutterstock, Inc.

© Cathleen A Clapper, 2011. Used under license from Shutterstock, Inc.

The tack refers to the equipment used on the horse for riding or other work. Typically in riding we use a saddle and bridle. The type of saddle further depends on the discipline or type of riding to be done, but is broadly categorized as either English (for English types of riding, to be discussed later) or Western saddles. Other disciplines may have more specialized saddles – like for racing or dressage. The saddle is placed on pads or blankets and is held in place with the use of a girth or cinch. The saddle is fitted to the horse as well as the rider.

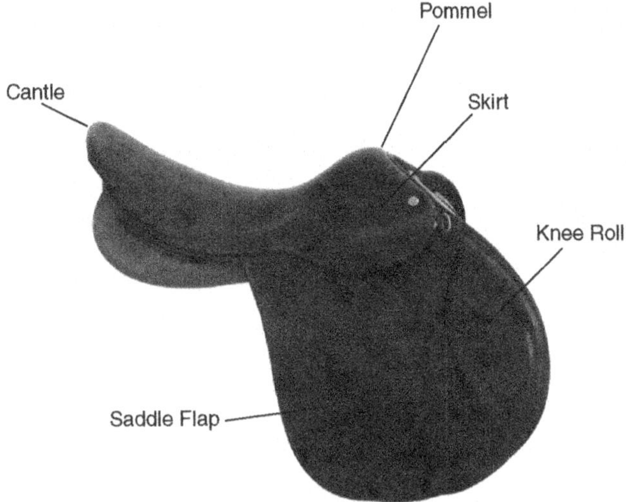

© marekuliasz, 2011. Used under license from Shutterstock, Inc.

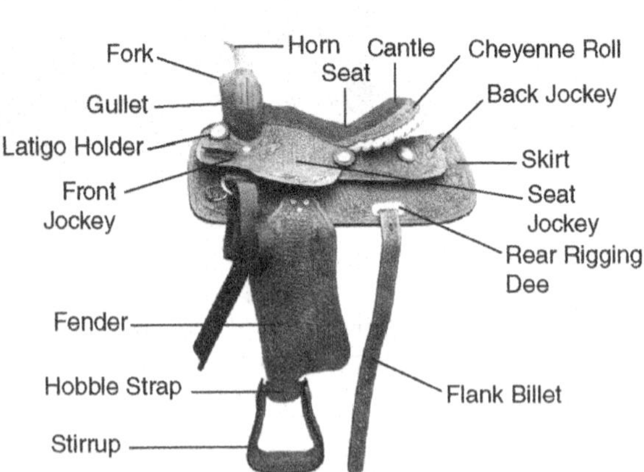

© Margo Harrison, 2011. Used under license from Shutterstock, Inc.

When mounting a horse, the rider traditionally mounts from the left side. This is based on riders of history having a sword on their left side (for ease of grasping with the dominant right hand). Mounting from the left would prevent the sword from being swung over the horse's backside.

The bridle is placed on the horse's head and has reins that the rider uses for control of the horse. Bridles either make use of a bit, or are bitless, which is usually called a hackamore. There are two main types of bits; curbs and snaffle bits. With a snaffle bit, the reins attach directly to the piece in the horse's mouth and apply pressure to the corners of the horse's mouth. For example, if you pull the right rein you will put pressure on the right side of the horse's mouth. With a curb bit, there are shanks at either side that the reins attach to. These allow the use of leverage where pressure is applied at the poll, the bars and the roof of the mouth, and primarily regulates the height the head is carried. A curb bit is used most frequently in Western styles of riding where neck reining (steering based on pressure of the reins on the neck) is common.

© Zuzule, 2011. Used under license from Shutterstock, Inc.

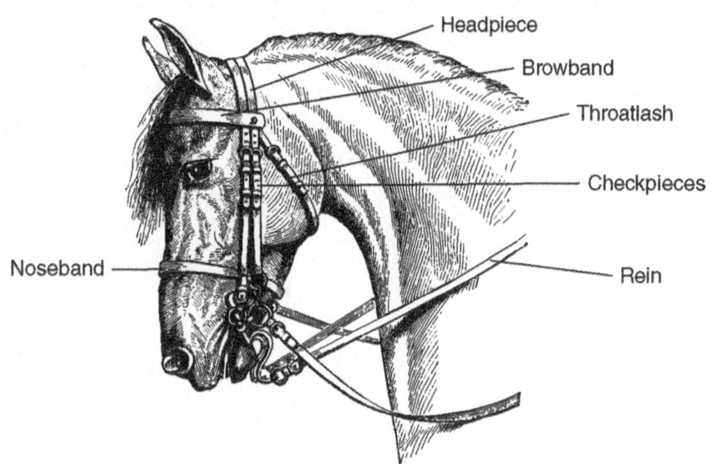

© Hein Nouwens, 2012. Used under license from Shutterstock, Inc.

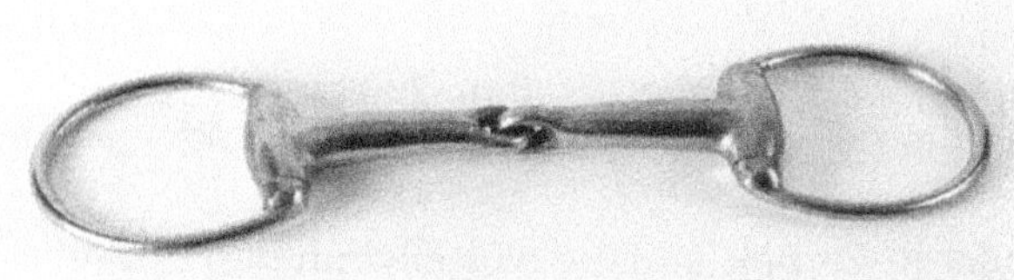

© Ad van Brunschot, 2011. Used under license from Shutterstock, Inc.

© Dani Simmonds, 2011. Used under license from Shutterstock, Inc.

© Sari ONeal, 2011. Used under license from Shutterstock, Inc.

Section 5.2 Equestrian Sports & Disciplines

Racing

Racing is a sport where the fastest horse wins. There are many different types of racing; Thoroughbred racing (which uses the breed Thoroughbred horses - and includes both flat racing and steeplechases), Standardbred racing (which uses Standardbred breed horses) includes both pacers and trotters, Quarter horse racing (using the breed Quarter Horse) and endurance racing (which typically uses Arabian breed horses, but may use Arabian crosses or mixes).

In Thoroughbred flat racing, Thoroughbred horses gallop distances of 5 furlongs (1 furlong = 220 yards) to 2 miles. A jockey rides the horse with a small lightweight racing saddle, and the horses are often handicapped with additional weight to even the field. There are three "big" Thoroughbred races, both in England (where such racing was popularized) and the U.S. In England these Triple Crown races are the 2,000 Guineas, the Epsom Derby and the St. Lager Stakes, and have been held since 1814. In the U.S., the Triple Crown includes the Kentucky Derby (first run in 1875), the Preakness Stakes and the Belmont. Thoroughbred horse racing is also very popular in France, Ireland, the middle East (Dubai in particular), Japan, Hong Kong, South Africa and Australia. Thoroughbred racing is strongly supported by gambling, which can contribute towards the prize money. In the US, we use the Pari-Mutuel betting system, which was developed by Pierre Oller. This system allows for a fixed percentage of the wager to go to the track for use in operation costs, the purses, etc. The remainder determines the payoff to the winner (s) of the bet. The projected payoff is based on the odds, which are established based on the number of bets placed on it. There are also several types of wagers that can be made, for example to win, place or show, or combination wagers such as exactas. There are also many types of Thoroughbred races, including handicap races, maiden races (for a horse that has never won a race before), etc.

Thoroughbred records as of publication:

- 6 furlongs – 1:06:49 ~40 mph (Twin Sparks, 2009)
- 1 mile – 1:32 1/5 ~ 38 mph (Dr. Fager, 1968)(Narjan, 2003)
- 1.5 mile – 2:24:00 ~ 37.5 mph (Secretariat, 1973)

Steeplechase races are very popular in Europe, and are typically run over 2-4 miles with obstacles or fences the horses must jump. Steeplechasing got its name from horses and riders navigating across fields (and fences) from church steeple to church steeple. Some of the "big" races include the Grand National in England (first race was 1839) and the U.S. Grand National, which is held at Belmont Park. There are also races with smaller obstacles called Hurdles or Point to Points, the latter of which is popular with amateurs.

Standardbred racing is also called harness racing and is very popular in the Northeastern U.S. and in Ontario, Canada as well as in Scandinavia. In this sport Standardbred horses are driven by drivers in specialized carts called sulkies. Depending on the horse's breeding, the horse either trots or paces. There are two Triple Crowns of Harness Racing; for Trotters there is the Hambletonian, the Yonkers Trot and the Kentucky Futurity; for Pacers there is the Cane Pace, Messenger Stakes and the Little Brown Jug. Horses race one mile and start with a moving gate (rather than in fixed gates such as in Thoroughbred racing). Horses also wear specialized equipment (in addition to the harness) including hobbles which help keep the horses in their desired gait.

Some records for Standardbred racing are:

- Pace – 1:46.0 (tied Always be Miki 2016; Lather Up, 2019) ~33.9mph

- Trot – 1:48.4 (Homicide Hunter, 2018) ~ 32.4pmh

Quarter Horse Racing is when American Quarter Horses race a quarter of a mile. This is a short, fast (45-55 mph; run in less than 20 seconds) race, where the winner is often the first horse out of the gate. Quarter horses are ridden by jockeys with similar lightweight racing saddles as in Thoroughbred racing.

Endurance racing is when horses are ridden over many miles, upwards of 50-100 miles. The Tevis Cup is a famous endurance ride of 100 miles over highly variable and often rocky terrain. Other races, such as those in the Middle East, may be on flatter sand terrain. Horses average about 12 mph and will walk, trot or canter throughout the ride. Many endurance races are run with loops, with a center base where horses can come back for rests. There are several forced rest stops in which horses are offered food and water, and are checked thoroughly by veterinarians. Arabian horses excel at endurance racing.

Many other breeds of horses are raced for multiple distances around the world.

Equitation – Ridden – Sports

When horses are ridden, particularly within the United States, there are two broad categories of riding; English and Western. There are also many different styles of riding that may not particularly fit these two categories, but for the purposes here we will consider them.

English or Hunt-seat riding originated in England and Europe and includes disciplines such as show jumping, foxhunting, dressage and equitation. Features of English style riding:

- Uses English saddle
- Typically the rider uses two hands on reins and has direct contact with the horse's mouth
- Gaits are called the Trot and Canter
- English riders traditionally posts (rises) the trot (except in dressage where they sit the trot)
- Snaffle bit is the most common
- Attire is more "formal" (wearing a hard hat, breeches and tall boots)

Western riding was developed in the United States in disciplines such as rodeo events, trails and western pleasure. Features of Western style riding:

- Uses Western saddle
- Reins held in one hand and neck reining with loose contact is used
- Gaits are called the Jog and Lope (though the footfall gait pattern is the same as the trot and canter)
- Western riders traditionally sit the trot
- Curb bit is the most common
- Attire is more "casual" – jeans, cowboy hat, cowboy boots, chaps (except in some disciplines)

English Disciplines

Showjumping is a test of height and speed. In many showjumping classes, there is a first round in which riders must navigate a field of ~12-16 obstacles (with heights based on skill level) in a given time. If the horse and rider are "clean" the first round, it means they have successfully completed the course with no rails knocked off the fences, no refusals and no time faults for having gone over the time allowed. If more than one horse and rider team goes clean, there is a "jump off" round in which an abbreviated course is ridden. In the jump off, the fastest "clean" round wins. There are also variations of showjumping classes such as speed classes (where time is added if a rail is knocked off a fence and the fastest time wins), or the Puissance, which is essentially the high jump, where the last fence of two to six obstacles gets higher and higher. The current record is 7 ft, 10 inches.

Equitation classes are a judged event where the rider's position and ability is judged. The focus is on the rider vs. the horse, however it is up to the rider to ride the horse effectively. Equitation classes may be held over fences (jumping), or on the flat (no jumps, just walk, trot and canter).

Hunter classes are also judged, though in these classes the focus is on the horse's jumping style and way of going. Conformation of the horse and its movement may also be taken into consideration. Typically in a hunter division there are 3-4 over-fences classes, and one "under-saddle" flat class (no jumps) where the horse's movement at the walk, trot, canter (and perhaps hand-gallop) are judged.

Dressage is a discipline named for the French term for training. It is based on a series of movements performed in a set pattern that is established for a given competition and skill level (except for the freestyle events where the riders develop their own "routine"). The movements executed by the horse and rider are the walk, trot and canter, which are also performed at collected and extended paces. The half pass, in which the horse moves laterally (sideways) across the arena, may be conducted at both the trot and canter. A piaffe is a trot in place and a passage is a slow motion trot. Tempi changes are flying lead changes, performed after only one or two strides each (when performed at each stride the horse almost skips from its right lead to left and back again). Horses may also perform a pirouette. These movements are performed in an arena, and each movement is executed at a specific position within the arena which are usually labeled by letters. The horse and rider are judged based on the excellence of the movement, including how the rider cues the horse to perform the task, with minimally obvious cues being rewarded. Each movement and transition is scored, with an overall score based out of 100. At the upper elite level of the sport, riders may design their own course pattern, and often set it to music, called the free-style or "Kur" event.

Fox hunting involves the tracking and chase of the fox, though true fox hunting resulting in the killing of the animal is illegal in many places. Instead, drag or trail hunting is popular, where an object is dragged over the ground to lay a scent to be followed. The field of riders involved in a fox hunt include many officials such as the Master of the Fox Hounds, the huntsmen, the kennelmen and the Whippers-In.

Eventing (sometimes called "combined training") is a sport involving elements of dressage, a cross-country jumping phase and stadium jumping (similar to show jumping). At the upper levels, the three phases are held over three (where the name three-day eventing comes from) or four days, but they may be held on one or two days at the lower levels. The cross-country phase requires the horse and rider to navigate a long course (2-4 miles long) with 20-40 obstacles (jumps) of varying height and difficulty. Often there are ditches, ponds and banks to be navigated as well. Historically in the upper levels of the sport there was also two "roads and track" phases and a steeplechase held along with the cross-country event to create a test of endurance. The difficulty of the dressage test and stadium jumping at the elite/Olympic level eventing competition is lower compared to the individual sports of dressage or showjumping. This is because one horse (typically a thorough-cross type of horse) is used for all three phases, and it would be (nearly?) impossible to compete at the elite level for all three sports.

Polo is an equestrian sport with four riders per team. Points are awarded to teams when they score by hitting a polo ball with a polo mallet into the goal, while mounted on their horse. Chukkers (also called chukkas) are typically 7 minutes each and are similar to a "period" in hockey. Between chukkers, the riders change mounts to allow the polo ponies a rest. There are 4-8 chukkers per match, so a team player would require at least 2 polo ponies (for a 4 chukker match), and ponies can only play 2 chukkers per game. After each goal, the ends change. Though called ponies, polo ponies are usually full sized horses, of various breeds (though Thoroughbreds and Thoroughbred-crosses are common).

Saddle Seat is a style of riding using American Saddlebreds, Tennessee Walking Horses, Arabians, Morgans and Hackneys, where the horse's high-stepping gaits are displayed. It is a judged event with classes including three-gaited (American Saddlebreds), five-gaited (includes the rack and a slow-gait), park, pleasure and equitation. The saddles and riders position in these events are slightly different from more traditional "English" styles of riding.

Western Disciplines

Rodeo events are the most well-known of the Western disciplines, and includes sports such as barrel racing, bronc riding, roping, cutting and penning. Reining, trail classes and show classes such as western pleasure are also examples of Western disciplines.

In Barrel racing, a horse and rider team follow a cloverleaf pattern around 3 barrels for the fastest time.

In Bronc riding, a rider attempts to stay on a bronc for 8 seconds and is scored out of 100, where 50 points are based on the horse and 50 points based on the rider.

In roping, the goal is to capture and restrain a calf using a lasso during a timed event.

In team roping a steer is used, and the two rider/horse team includes a header and a heeler to secure the head and the feet of the animal, respectively.

In cutting and penning, the goal is to separate an animal from its herd. In cutting typically one cuts a cow (or other bovine) and keeps it away from the herd, while in penning the cow is moved into a pen, and is usually done as a team event. Cutting horses in particular show "cow sense" where they seem to anticipate the movement of the animal.

Reining demonstrates the athletic ability of a ranch-type of horse. A horse and rider team run one of 10 approved patterns that includes small slow circles, large fast circles, flying changes, roll backs over the hocks, 360 degree spins (in place), and sliding stops.

 Trails classes are a judged event where the horse and rider team must navigate obstacles that might be encountered while on a trail, such as gates, bridges and rails.

Western pleasure is a judged event based on the rideability of the horse, manners and temperament and way of going. Horses are judged at the walk, jog and lope, and sometimes the gallop. There are also western equitation and western dressage classes, somewhat similar to those described for English styles of riding.

Gymkhana events are games on horseback typically designed for younger riders. Events include barrels, pole bending, relays and keyhole games.

There are many types of competition where the horse is driven, rather than ridden (in addition to harness racing).

Combined driving is an event where horses compete in teams (typically one, two or four horses) pulling carts. Depending on the number of horses, there may be up to four people in the cart, with roles as drivers, grooms and navigators. There are three phases to this type of competition; dressage, cross country marathon and obstacle cone driving. In the dressage phase, horses are judged based on presentation and their movement in an arena, similar to traditional dressage. Movements include circles, figure eights and diagonals at the walk, trot and canter. The marathon event is when the team navigates hazards such as water, twists and turns around objects or steep hills. The cone driving section is a test of accuracy where the driver navigates a course of cones with a ball on top. Thus, the drivers must be fast and accurate to complete the course within a time allowed and without knocking over a ball or cone.

Draft horse competitions include those where a team of horses is hitched to a large carriage and is judged on the horse's performance and manners. Draft horse pull competitions are those where horses (single horse or a team of two horses) must pull a given weight a certain distance, with the weight increasing in each round.

Carriage driving events are those where fine carriages or carts are pulled by a single horse, or two or four-horse teams. They are judged on the presentation of the horse and carriage and manners and performance of the horses.

Fine Harness classes are similar to carriage events though there is an emphasis on the show of the gait and the high stepping action of the horses.

Roadster events are those where horses are ponies are dressed as for racing (in silks pulling sulkies) but are judged rather than raced.

Vaulting is a sport similar to gymnastics, but on horseback. Competitors may be individuals, pairs or teams. The horse canters in a 15 m circle on a line and the athletes mount the horse from the ground and perform movements such as handstands, flips or lifts on board the horse.

Section 5.3 Other Equine Uses

Horses, in particular miniature horses, are becoming a common guide animal. The Guide Horse Foundation suggests that because horses have a longer lifespan than dogs, they may be more suitable guide animals.

http://guidehorse.org/

It is becoming increasingly well known that horses make good therapy animals. Horses can be used for physical therapy for those with physical disabilities to help these people build core strength. Horses can also give some freedom of movement for those confined to a wheelchair. Those with mental or behavioral disabilities can form close bonds with horses, giving them emotional support and confidence. Horses have shown to improve the well-being of those with autism, PTSD and several types of behavioral disorders.

Horses are still used in many aspects of ranch work. They are used as cow horses to help maintain or muster or round up the herd. They are still used in farming for hitch and plow work.

Horses, and in particular mules and donkeys are still used to transport materials on their backs. Alternatively they may be harnessed to pull material behind them, such as for logging or mining.

Mounted police are still found in many cities, to facilitate with crowd control at events, and in congested cities. They may also be used for border patrol. They are also often used for search and rescue operations, particularly in remote areas where vehicles may not be practical.

The Royal Canadian Mounted Police is the national police force of Canada, and while historically the officers were mounted on horseback, they still maintain the history through the RCMP Musical Ride.

CHAPTER 6

Equine Anatomy

The objectives of this section are to learn about the basic systems and structures (bone, muscle) of the horse

Section 6.1

The body is organized as tissues, organs and systems. A tissue is defined as a group of similar cells. When tissues function together they form an organ, and a group of organs with a similar overall function make up a system. Several systems will be discussed in more detail later; the skeletal system (section 6.2), the muscular system (section 6.3), the digestive system (section 10.3) and the reproductive system (section 10.2). The others will be discussed briefly here.

The **skin and coat** of the horse functions to protect the internal systems from injury and the body from dehydration, for vitamin D synthesis and to assist with temperature regulation. A vitamin D precursor is formed in the skin when the skin is exposed to the UV rays of the sunlight. Horses are very efficient at sweating and dissipate a great deal of their heat through sweat. Some horses have a condition called anhidrosis where they can't sweat, and these horses have difficulties performing athletic activities because of it. The skin and coat is also very important for sensation (touch).

The **urinary system** functions to remove wastes from the blood and to expel them from the body via the urine. It also facilitates the maintenance of blood pH. About 1000 milliliters of blood is filtered by the kidneys per minute.

The **respiratory system** functions to facilitate the exchange of oxygen and carbon dioxide. There are two main components of respiration; external respiration (breathing) and internal respiration (oxygen use at the cell). The structures of the respiratory system include the upper airway, consisting of the nostrils, nasal cavity, nasal septum, turbinate bones, sinuses, pharynx, soft palate, epiglottis, larynx and trachea. The lower airway consists of the trachea and lungs, with the bronchi, bronchioles and alveoli, as well as the diaphragm. The lungs have a total capacity of about 55 liters.

Breathing is controlled by the respiratory center in the medulla oblongata of the brain that detects changes in oxygen, carbon dioxide and pH. Breathing is also under some voluntary control, such as for phonation (vocalization) and during efforts such as parturition (giving birth) and defecation. When horses run at the gallop, respiration is coupled to locomotion, such that a horse will take a breath with each stride. Horses will inhale during the suspension phase of a gallop.

Roaring is a condition called Laryngeal Hemiplegia which is a result of paralysis of left arytenoid cartilage. This results in the cartilage flapping in the airway causing a roaring noise during exercise, and potentially inhibiting airflow. There is a surgical treatment called "tie back" surgery that is very effective.

<http://evrp.lsu.edu/healthtips/LaryngealHemiplegia.htm>

Exercise Induced Pulmonary Hemorrhage (EIPH) is condition seen in racehorses or other highly athletic animals. With intense exercise, the pressure within pulmonary capillaries increases to a point that can rupture the vessel walls, causing blood to leak into the alveolar sac and can exit via the nostrils (when blood is observed at the nostrils it is called epistaxis). There may be other physiologic causes as well.

Heaves is also called recurrent airway obstruction (RAO) or Chronic obstructive pulmonary disease (COPD) and is very similar to the condition asthma in humans. It is caused by irritants to the respiratory tract such as molds and dust, and often removal of such irritants by having the horse live outdoors is good management practice.

The **circulatory system** functions to supply the body with nutrients and remove wastes. It involves the heart, arteries (carry blood away from the heart), veins (carry blood too the heart) and capillaries (that facilitate gas and nutrient exchange at the tissues and lungs). Blood is pumped through the arteries and capillaries thanks to the pumping action of the heart. Blood is returned to the heart thanks to one-way valves on the veins, muscular contractions facilitating movement of blood through the veins and by the action of the digital cushion in the horse's hooves that help push blood back up to the heart. The horse's heart weighs about 8-10 pounds.

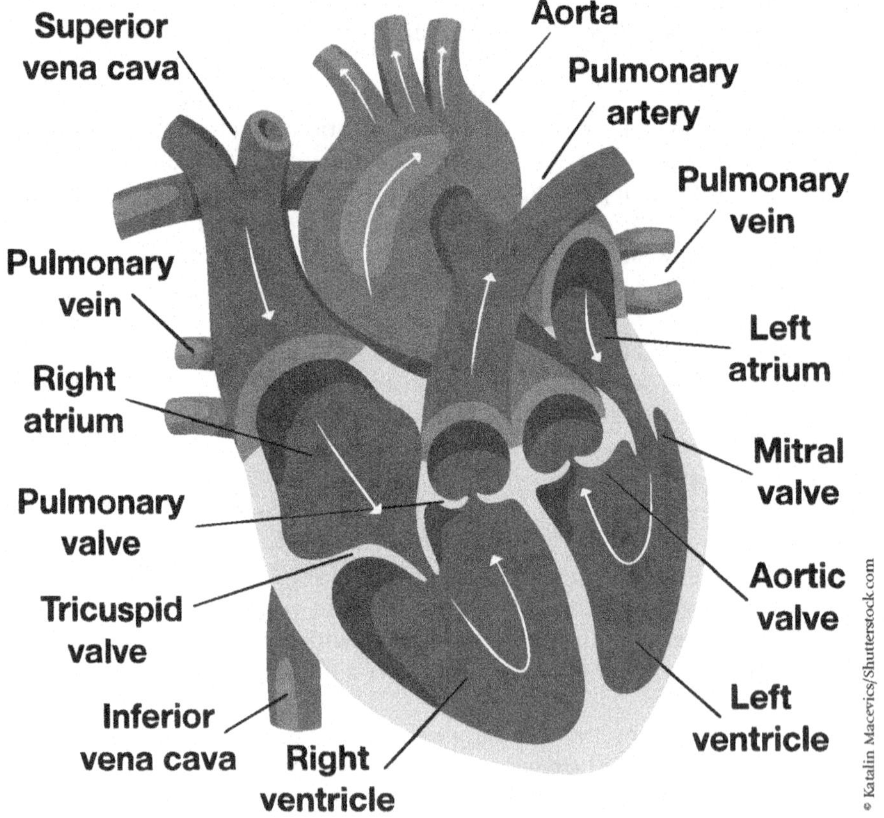

Blood accounts for approximately 9% of the horse's body weight, with approximately 80% of the blood in the systemic circulation, and 20% in the pulmonary circulation. Most of the blood is located within the veins. Within the blood there are red blood cells, white blood cells, platelets and plasma. The hematocrit is the percentage of red blood cells in the blood. With stress, the amount of red blood cells increase, thus increasing the hematocrit, thanks to the horse's spleen that functions to store and release red blood cells as needed.

Blood samples are easily taken from a horse from their jugular vein. Specialized collection tubes may contain anticoagulants for the collection of plasma or no anticoagulant for the collection of serum.

The control systems include the nervous system and the endocrine system. Both function to maintain homeostasis within the body. They respond to stimuli and react accordingly. The nervous system includes the brain, spinal cord and nerves. The endocrine system includes the glands (hypothalamus, pituitary, adrenal glands, thyroid, parathyroid, thymus, pancreas, ovaries/testes) and hormones.

The Senses

- **Sight**
- **Hearing**
- **Smell**
- **Taste**
- **Touch**

Horses have a wide range of vision thanks to the placement of their eyes on the side of their heads. However, they have mostly monocular vision (field seen with only one eye) and a small area of binocular vision (field seen with two eyes, which is required for depth perception). They also have a blind spot immediately in front of them (small, right between their eyes and close to their face) and immediately behind them. Horses have good night vision thanks to their tapetum lucidum which helps to reflect light within the eye. The horse's eyeball is flattened in shape which allows the horse to focus both on near and far objects easily.

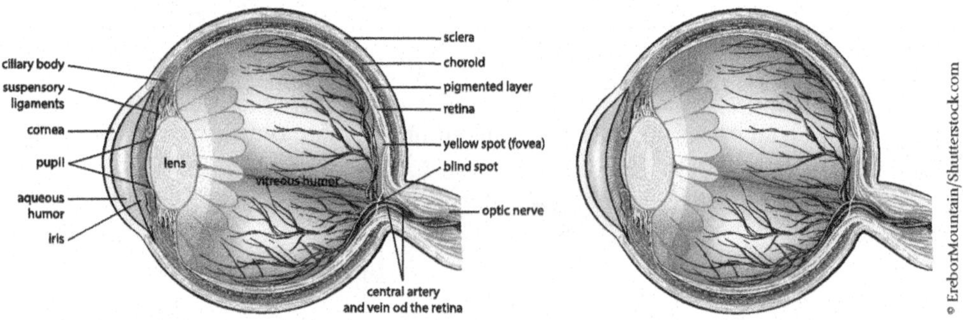

Horses have excellent hearing, in part thanks to their ears that can rotate 180 degrees. Sound waves cause the eardrum to vibrate, which is converted to electrical nervous impulses.

Taste is highly linked to smell, and horses tend to prefer sweet tastes to others.

Smell is important for the selection of food, detection of predators and for communication. The organ of Jacobson (also called the vomeronasal organ) is a specialized collection of olfactory cells that are exposed with the Flehmen response (when a horse raises his upper lip), and that are able to detect chemicals called pheromones.

Touch requires nerve endings in the skin that detect changes in pressure and transmit the signals to the brain. Horses are particularly sensitive to touch along the ribs. Horses use touch for communication, such as through the action of mutual grooming.

Section 6.2 Skeletal System

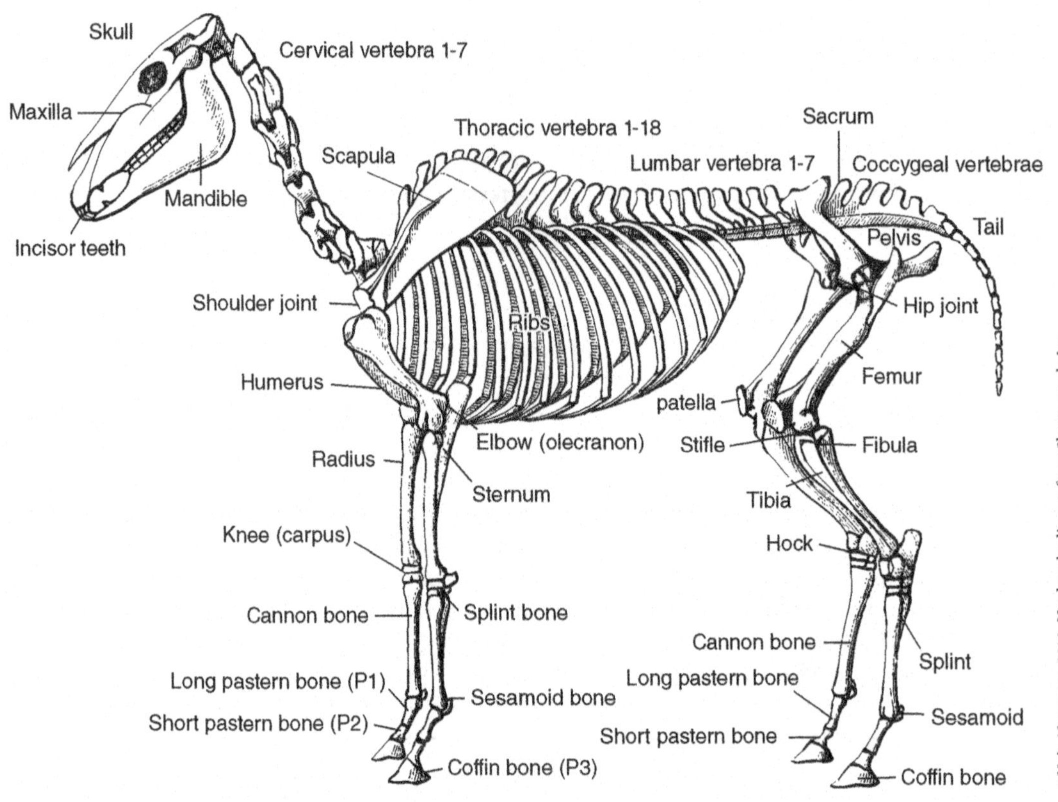

Parts of the Skeleton

The axial skeletal bones are those that run along the main central axis of the body.

- Skull
  - Frontal bone
  - Nasal bone
  - Temporal bone
  - Parietal bone
- Maxilla
- Mandible
- Vertebrae
  - Cervical vertebrae: 7
  - Thoracic vertebrae: 18
    - Spinous processes = withers
  - Lumbar vertebrae: 6
    - Transverse processes
  - Sacral vertebrae: 5
    - Fused = sacrum
    - Pelvic bones attach at sacro-iliac joint
  - Coccygeal vertebrae: 15-21 (average 18)
  - Tail
- Ribs
  - 8 true ribs (attach to sternum)
  - 10 false ribs (not connected to sternum)
- Sternum
  - Breastbone

The appendicular skeleton is attached to the axial skeleton at the pelvic girdle and the pectoral girdle. Unlike humans, horses do not have a collarbone, so the limbs are attached to the axial skeleton by muscles, tendons and ligaments.

Forelimb

- Scapula
- Humerus
- Radius
- Ulna
- Carpus (carpal bones)
  - Knee
- Metacarpals
  - Cannon bone (Metacarpal III)
  - Splint bones (2) (Metacarpals II and IV)
- Phalanges
  - Long pastern bone (P1)
  - Short pastern bone (P2)
  - Pedal/Coffin bone (P3)
- Sesamoid bones
  - Proximal sesamoid bones
    - Behind fetlock joint
  - Distal sesamoid bone
    - Navicular bone

Hindlimb

- Pelvis
  - Ilium
  - Ischium
  - Pubis
- Femur
- Patella
  - Sesamoid bone
  - Knee cap
- Tibia
- Fibula
  - Small (vestigial)
- Tarsus
  - Hock
- Metatarsals
  - Cannon and splint bones
- Phalanges
  - Long pastern bone (P1)
  - Short pastern bone (P2)
  - Pedal/Coffin bone (P3)
- Sesamoid bones
  - Proximal sesamoid bones
    - Behind fetlock joint
  - Distal sesamoid bone
    - Navicular bone

Joints are the union of bone and bone. The joint capsule functions to cushion the bones, using synovial fluid and articular cartilage. Main equine joints include the shoulder, elbow, knee, fetlock, pastern, coffin, hip, stifle and hock joints. Horses may develop a condition called arthritis, which is the inflammation of the joint.

Section 6.3 Muscular System

There are three types of muscle; cardiac (found in the heart), smooth (found in arterial walls and the digestive tract) and skeletal muscle (allows for the movement of limbs.

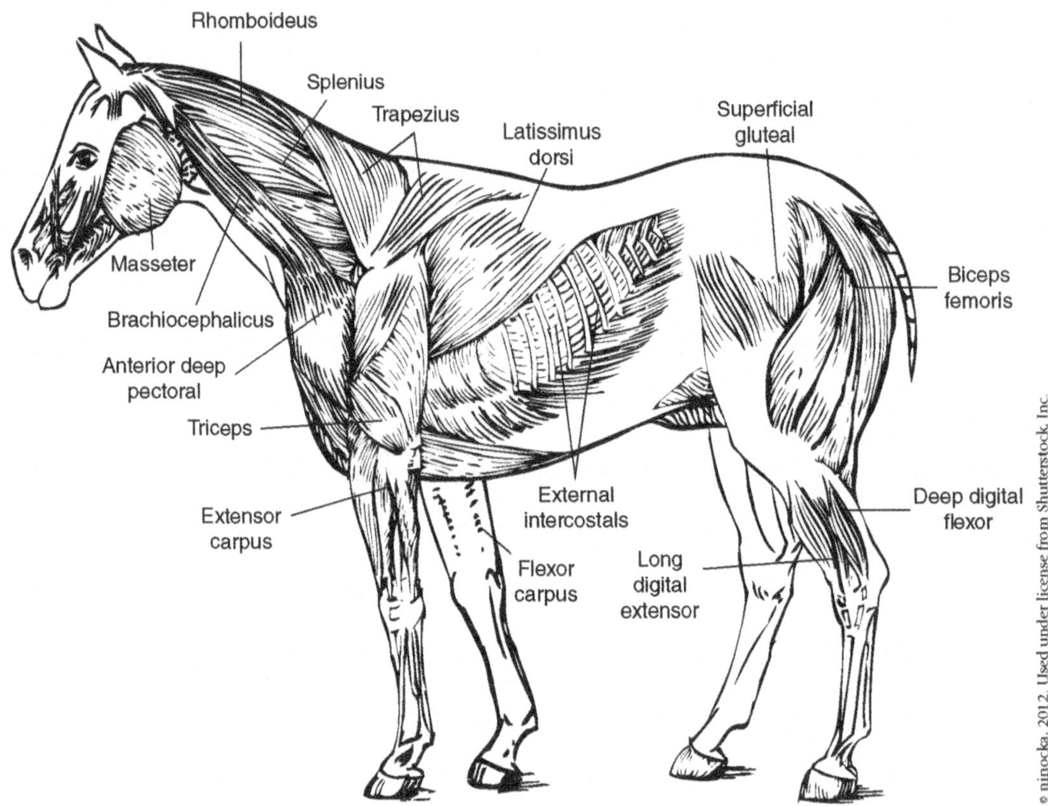

Key Muscles of the Horse

- Head
  - Masseter
- Neck
  - Splenius
  - Rhomboids
  - Brachiocephalic
  - Sternocephalicus
- Muscles of the Horse
- Forelimb
  - Deltoid
  - Triceps
  - Pectorals
- Trunk
  - Trapezius
  - Latissimus dorsi
  - Intercostals
  - Abdominal muscles
- Muscles of the Horse
- Hindlimb
  - Gluteals
  - Quadriceps
  - Hamstrings
  - Gastrocnemius (calf muscle)

Horses may develop muscle conditions also known as myopathies. Many of these have symptoms commonly called "tying up" where the muscles become very stiff and the horse is reluctant to work.

Recurrent Exertional Rhabdomyolysis is a disease that is commonly observed in racing horses, such as Thoroughbreds, Standardbreds and Arabians.

Polysaccharide Storage Myopathy (PSSM) is a disease that is associated with abnormal glycogen storage in the muscle. There appear to be several types of these disorders including glycogen branching enzyme deficiencies. Breeds that have been identified to have these diseases include Quarter Horses and related stock breeds. Another type of glycogen disease also appears in warmbloods.

Hyperkalemic Periodic Paralysis is a disease that is found in Quarter horses that have the stallion Impressive in his bloodline. This is a genetic disease that will be discussed further in that section.

Further Reading

- Equine Science
  - Pilliner and Davies, 2004
  - ISBN: 1405119446
- Horse Anatomy
  - Green, 2006
  - ISBN: 0486448134

CHAPTER 7

The Hoof & Lower Limb

The objectives of this section are to learn about the parts of the hoof and lower limb, and common problems associated with them.

Section 7.1 – The Hoof

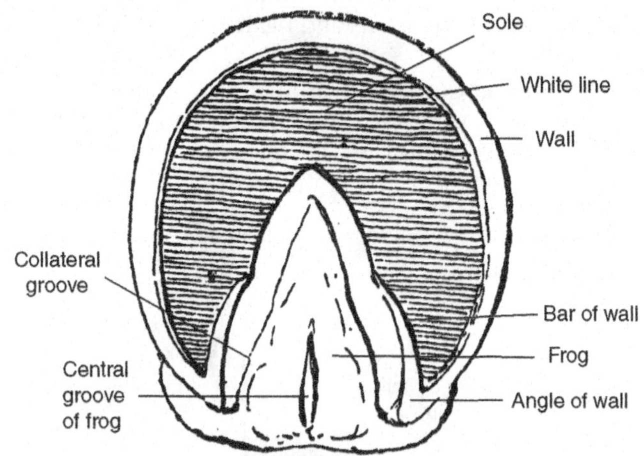

© Morphart Creations inc., 2012. Used under license from Shutterstock, Inc.

The hoof functions to protect the internal structures of the hoof, to support the weight of the horse and reduce concussion associated with movement, and to prevent slipping. The sole has a suction effect on surfaces, and the toe also facilitates cutting into footing to help prevent slippage. The hoof also facilitates circulation, by promoting venous return of blood back up towards the heart. The pressure and weight in the hoof compresses the frog and digital cushion, and when weight is released these structures spring back and helps to push blood up and away from the extremities.

The external hoof wall of the hoof grows from the coronary band, similar to the human fingernail growing from the cuticle. The wall is longest at the toe area and shortest at the heel, and intermediate at the sides or quarters. The wall is made of cornified tissue to give it sufficient strength to protect the internal structures and to reduce concussion. The wall is also insensitive, similar to our own nails, and can be trimmed if it grows too long.

The frog is a v-shaped structure on the bottom of the hoof with the point towards the middle of the sole. The frog assists the digital cushion in helping to pump blood back to the heart and plays a role in reducing concussion due to its tough spongy texture.

The sole is the base part of the horse's hoof. It comes into contact with the ground, particularly in horses that are not wearing shoes. The extent to which it contacts the ground affects its texture, being tougher with regular ground contact.

The inner part of the wall is called the laminae or corium, which is highly sensitive. This is a highly vascular tissue between the external wall and the coffin bone, and provides a contact surface and grip between the coffin bone, and the external hoof wall. Damage or inflammation to the laminae is extremely painful.

Also inside the hoof is the digital cushion, a spongy tissue that helps with circulation and concussion. In addition to the coffin bone are the lateral cartilages, the navicular bone and the tendons and ligaments of the lower limb.

Section 7.2 The Lower Limb

The lower limb of the horse includes the bones (cannon bone, splint bones, long pastern bone, short pastern bone, coffin bone, navicular bone and the proximal sesamoid bones), as well as the tendons and ligaments.

The limbs support the weight of a large animal into a small area. Approximately 60% of the horse's weight is supported by the front end, thus for a 1000 lb horse, about 300 lbs of weight is supported by each front leg.

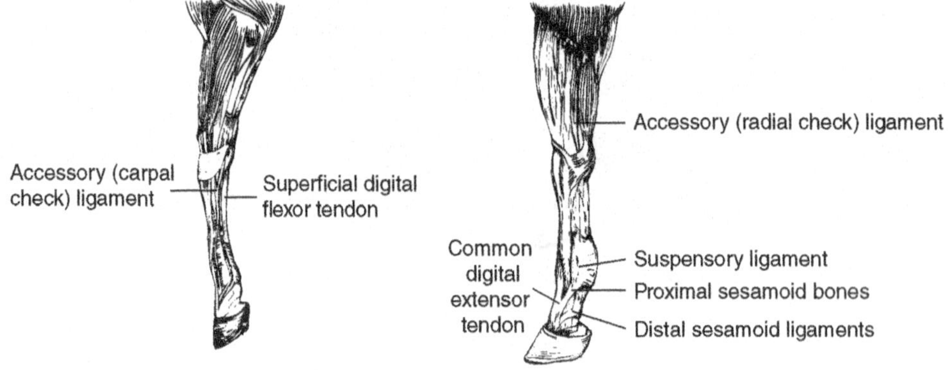

© ninocka, 2012. Used under license from Shutterstock, Inc.

There are no muscles below the knee or the hock, so movement (flexion and extension of the fetlock, pastern and coffin joints) is as a result of tendons and ligaments acting on these bones. Tendons attach muscle to bone, so when a muscle contracts, it pulls the tendon which in turn moves the bone. This is similar to strings attached to a marionette doll, when pulled the move the doll's limbs. Tendons are made of collagenous fascicles, plus tendon sheaths around them for protection.

The key extensor tendons (that increase the angle of a joint) are the common digital extensor tendon (runs down front of cannon from knee to coffin bone) and the lateral digital extensor tendon (runs from side of knee to long pastern).

The key flexor tendons are the superficial digital flexor tendon (begins above the knee, splits below fetlock and attaches to long and short pastern) and the deep digital flexor tendon (begins above the knee, attaches at pedal bone underneath navicular bone).

Ligaments attach bone to bone. The check ligament attaches the deep digital flexor tendon to the back of the knee. The suspensory ligament is actually quite elastic and is between the deep digital flexor tendon and the cannon bone. The suspensory ligament is very important in helping to provide support through the fetlock and transmitting concussive energy through the hoof. If you watch a horse canter, gallop or land after a fence, you will see how close the fetlock comes down to the ground; it is the suspensory ligament that prevents the fetlock from breaking down, and assists with springing the leg back up. The suspensory ligament is also important in the stay apparatus of the horse.

The stay apparatus allows the horse to rest when standing while using minimal muscle energy. The horse's limbs have several angles (like between the shoulder blade to the humerus and forearm), which together with muscles and tendons to create a system of support for the limb. The angles work similar to rod that is angled under a shelving unit to give support for the objects on it. The angle through the fetlock to the pastern is strongly supported by the suspensory ligament. In the hind limb, there is another system, called the locking stifle mechanism. When a horse locks its patella, the stifle and hock lock in place to create a strong support, and this allows the other leg to rest.

Section 7.3 – Common Hoof Problems

Common hoof problems include bruises, corns, wall cracks and cracked heels. Thrush is a bacterial infection of the frog resulting in a black and crumbly frog that occurs when the frog is not kept clean in damp and wet bedding.

Navicular disease is an inflammation or degeneration of the navicular bone that results in lameness. It can be caused by strain, overuse or poor conformation such as upright pasterns or low heels. Depending on its severity it can be managed with drugs, specialized shoeing or surgery.

Laminitis is the inflammation of the laminae in the hoof. This results in connectivity between the coffin bone and the hoof wall to weaken, which may result in downward rotation of the coffin bone. This sinking of the coffin bone is termed founder and may be a chronic condition. The inflammation is very painful (similar to having a fingernail ripped off a finger, but then having 300 lbs of weight pressed into it!). Typically laminitis affects both front feet, and the horse will attempt to keep weight off them resulting in him leaning back. Laminitis is a complicated condition that can be caused by diet or grain overload, metabolic conditions associated with obesity, trauma, concussion, uneven weight distribution in the limb resulting in poor circulation in the hoof, and excessive pasture intake.

Sidebone is the ossification of the lateral cartilages of the coffin bone that may cause lameness. Ringbone is a bony growth at the joints of either the long and short pastern bones (high ringbone) or the short pastern bone and the coffin bone (low ringbone).

Arthritis in any of the lower limb joints may be acute as a result of trauma or infection, or chronic as with degeneration of the joint capsule most commonly seen in older athletes. Specifically, carpitis is arthritis of the knee, and osselets is arthritis of the fetlock. Treatment for arthritis includes rest, pain medication, or joint therapy. Joints may be directly injected with hyaluronic acid, one of the constituents of joint fluid in effort to replenish the buffering ability of the joint capsule. Alternatively oral joint supplements may be consumed in effort to provide the compounds for the body to make new joint fluid, though research investigating the effectiveness of these supplements is conflicting.

Spavins typically refer to conditions affecting the hock area. Bone spavins are a bony growth at the hock that are a type of arthritis. Bog spavins are swellings at the hock joint that may be caused by an infection or strain. Blood spavin is a distended vein in the hock.

Splints are inflammation of the splint bones. Recall that metacarpal/tarsal II and IV are the splint bones and are on either side of the cannon bone. The inflammation may be caused by overuse or trauma and results in swelling, heat and lameness. Treatment for splints includes rest, anti-inflammatory medication or surgery.

A bowed tendon is tendonitis, a tear or inflammation of the tendon fibers causing them to be swollen and "bow". Most commonly we see a bow in the superficial digital flexor tendon as a result of strain or poor conformation. Treatments include rest, cold therapy and anti-inflammatories, though new therapies that have promise include stem cell and shockwave therapy.

Windgalls or windpuffs are swollen fetlocks, typically in the hind limbs. Often they are not serious and do not cause any lameness.

Further Reading

- Equine Science
  - Pilliner and Davies, 2004
  - ISBN: 1405119446
- Understanding Equine Hoof Care
  - Smith Thomas, 2006
  - ISBN: 1581501366

CHAPTER 8

Equine Health & Common Conditions

The objectives of this section are to learn about what normal equine health is, how to recognize signs of abnormal health and general health management. You will also learn some common equine defects, blemishes, injuries and unsoundnesses

Section 8.1 – Normal Health

It is very important to understand the features of a healthy horse in effort to better understand when a horse is unhealthy.

A normal healthy horse should be eating and drinking well, and have regular bowel movements and urinary habits. He should be responsive and alert. He should be in good condition, with ample fat coverage for his weight, his coat and hooves should be in good condition. The eyes, ears and nostrils should be clear. Another useful tool for assessing a horse's health is knowing his normal, healthy "TPR" – or temperature, pulse and respiration.

The horse's normal resting heart rate is 30-40 beats per minute (bpm). With exercise, the horse has the ability to achieve heart rates of 240 beats per minute! When a horse is born, their heart rates are usually higher (60-75 beats per minute) and these decrease over the first few years of life.

Measurement of a horse's heart rate can be done with a stethoscope or heart rate monitor. Also, a veterinarian can measure the electrical activity of the heart using an ECG (electrocardiogram). The horse's pulse rate can also be determined, which should be the same as the heart rate (as it represents the pulse of blood through an artery that was pushed by the heart's contraction). Major arteries can be found under the jaw bone, in the leg or under the tailbone, and can be used to take a horse's pulse rate.

A horse's normal respiratory rate is 8-16 breaths per minute. Typically this can be measured simply by watching the flank or nostrils for movement.

A horse's normal temperature is approximately 99 degrees Fahrenheit. However, it may range from 98-101, and anything over 102 degrees is reason to call a veterinarian.

In addition to TPR, a horse's circulation can be assessed easily, to determine how blood is flowing through the body. The horse's gums should be pink in color, with white or red/maroon gums being a sign of concern. Also, if pressure were applied to the gums, they should blanch (whiten) but return to the normal pink color within a few seconds. Delayed capillary refill is also cause for concern.

When any of these "normal" criteria for a healthy horse are off, a veterinarian may be consulted. Sometimes a horse owner may just sense that their horse is "off" or "ADR" (ain't doin' right) – and will check the horse's vitals and contact their veterinarian.

Section 8.2 Equine Health Management

A large part of general horse care and management is equine health management, or preventative care. This includes hoof care, dental care, vaccinations (section 9) and deworming practices (section 9).

Dental care is required by horses because their teeth continuously erupt throughout life, and due to their side to side chewing action. This may result in sharp points forming on their teeth, particularly on the cheek-sides of their molars. Equine veterinarians or specially trained dentists can "float" a horse's teeth, which is when they use a large file and file down those sharp points. Often a horse needs to be sedated or restrained for the procedure. Horse's should have their teeth checked at their annual exams, and may require floating every 1-2 years.

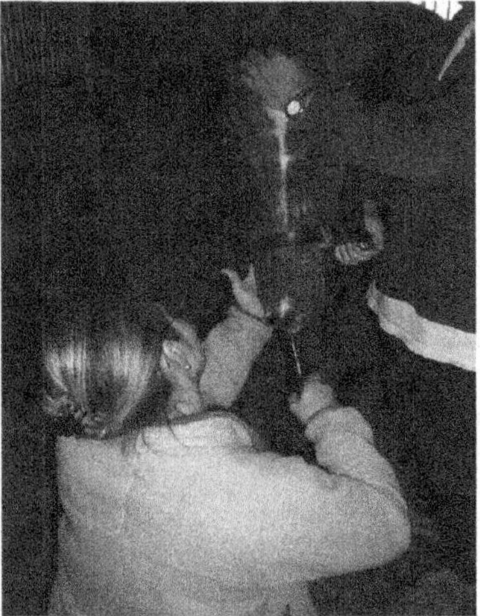

Photo courtesy of author

Equine hooves require regular (daily) cleaning with a hoof pick to remove debris and stones. Also, a horse requires regular farrier care, which usually involves trimming the wall and sole, and may also involve application of horse shoes. The hoof grows approximately ¼ – ½ inch per month, and therefore horses should have their hooves trimmed approximately every 6-8 weeks. A farrier will use a hoof pick to clean debris, a rasp to file down the hoof wall, a hoof knife to work away any sole, and nippers to help trim the wall, or to trim any horse-shoe nails.

Section 8.3 Equine Conditions

Despite the best preventative health care, horses may become afflicted with the following conditions. Other conditions that affect horses were discussed in the hoof anatomy section (thrush, laminitis, etc.) or may be discussed in future sections such as nutrition or genetics.

Eye conditions of the horse are considered medical emergencies, and your veterinarian should be contacted as soon as possible should you note anything irregular about your horse's eyes.

Corneal ulcer: Disruption to the cornea caused by trauma or infection

Conjunctivitis: Infection of the lining of the eye

Recurrent uveitis: AKA Moon blindness. Inflammation of the uvea (iris, ciliary body, choroid)

Lymphangitis: Infection of the lymph vessels

Fistula/Poll Evil: Bacterial infection of the withers (fistula) or poll (poll evil). Often caused by Brucella species and may be transmitted to humans. May also be caused by trauma.

Abscesses: Collection of pus under the skin causing swelling. Caused by foreign material working its way out or an infection

Girth Galls/Saddle Sores: Hair loss and irritation caused by friction due to poor fit or dirty conditions

Obesity: Body condition score >7, Overweight >6. Obesity is a significant problem in our horse industry. Results in exercise intolerance, heat intolerance, reproductive problems, metabolic problems and increased risk of laminitis. Any horse that has a body condition score greater than 6 should have its feeding program examined by a nutritionist.

Further Reading

- Complete Equine Veterinary Manual
  - Pavord, 2009
  - ISBN: 0715332791
- Horse Owner's Veterinary Handbook
  - Gore, Gore, Giffin and Adelman, 2008
  - 9780470126790

CHAPTER 9

Equine Disease and Preventative Care

The objectives of this section are to learn about equine diseases, parasites and preventative health care

Section 9.1 Non-Infectious Diseases

The majority of this chapter will discuss infectious disease or parasites, but there are several equine conditions that are non-infections. This means they are not caused by an infection of any kind, and are therefore not transmissible. These include genetic diseases, nutritional diseases and environmental diseases.

Genetic diseases will be discussed further in the Genetics section, but include conditions such as Hyperkalemic Periodic Paralysis (HYPP), lethal white foal syndrome and severe combined immunodeficiency disease.

Nutritional diseases include those caused by nutritional toxicity or deficiency (too much or too little of a required nutrient). Colic (which will be discussed further in the nutrition section), which is a digestive disturbance, can be a result of poor feeding management or parasitic infestation. There are several toxins that may be consumed inadvertently by the horse, either in its regular feed (such as botulism, which may be present in some forages such as haylage) or in some plants. Some highly toxic plants include hemlock, oleander and yew.

An example of an environmental disease includes heaves or COPD, as was discussed earlier.

Section 9.2 Infectious Diseases

https://aaep.org/infectious-disease-control/infectious-disease-guidelines

Infectious diseases are those that are caused by a virus, bacteria or even a parasite that may be transmitted to other animals either directly (horse to horse, usually via respiratory secretions) or via a vector (such as via a mosquito). Many of these diseases affect either the respiratory system or the neurological system. Relatively common equine infectious diseases include; tetanus, influenza, equine encephalomyelitis, rhinopneumonitis, equine viral arteritis, strangles, Potomac horse fever, rabies, West Nile virus and equine protozoal myeloencephalitis (EPM). Many diseases have vaccines that may prevent them. Some diseases are more serious (deadly) so it is recommended that all horses get those vaccines each year (called CORE vaccines). Other diseases may only pose a threat to a horse under some circumstances that are considered "at risk" for the disease (such as being in an "open" herd or area where it might come into contact with new or unknown horses; or horses that may be used for breeding).

Tetanus is also called lockjaw and is caused by a toxin produced by the bacterium *Clostridium tetani*, which is commonly found in the soil. Tetanus affects neuromuscular responses and therefore neurologic symptoms such as abnormal contraction of skeletal muscles are common symptoms. Horses in the early stages of tetanus may also show a prolapse of the third eyelid. Tetanus can be fatal if not caught early and treated. There is a CORE vaccine available to help prevent tetanus in horses.

Rabies is caused by a rhabdovirus that affects the nervous system by causing inflammation of the brain. It is transmitted from secretions from an infected animal, usually via a bite and saliva. Initial symptoms include fever, headache and mania, and can be fatal within a few hours, though it may take several days for the disease to set in. It is a zoonotic disease, which means that it can be transmitted to humans. There is a CORE vaccine available to help prevent rabies in horses.

https://epi.dph.ncdhhs.gov/cd/rabies/control.html

Equine Encephalomyelitis is also called Sleeping Sickness, due to its symptoms of droopy ears, lethargy and paralysis as a result of inflammation of the central nervous system. It is caused by an arbovirus, of which there are 3 strains: Eastern (EEE), Western (WEE) and Venezuelan (VEE). It is transmitted by vectors (mosquitoes), with birds and mammals serving as reservoirs. There are CORE vaccines available to help prevent the different strains of equine encephalomyelitis.

https://epi.dph.ncdhhs.gov/cd/diseases/eee.html

Equine Protozoal Myeloencephalitis (EPM) is caused by a parasite (protozoa) called Sarcocystis neurona. The parasite infects the central nervous system so the symptoms are neurological. It is often difficult to diagnose in live animals in part because the symptoms may be similar to other neurologic diseases (EEE, West Nile Virus). The parasite has a life cycle that involves a definitive host, the opossum, and other intermediate hosts; cats, raccoons, sea otters, armadillos, skunks. The horse is a dead-end host that is infected through feed or water contaminated with opossum feces. There is NO vaccine available to prevent EPM, and it can be difficult (and expensive) to treat.

Equine West Nile Virus is a neurologic disease that is spread by mosquitoes. Prevention is available through CORE vaccination and by decreasing exposure to mosquitoes, by decreasing breeding sites and using repellents.

Equine Influenza is caused by the influenza virus, which, similar to human influenza, is at risk for mutation and resistance. Symptoms of influenza infection include fever, cough and nasal discharge. It has a 100% infection rate, which means that if a horse is exposed and not vaccinated, it will develop symptoms. There is a vaccine which is considered a RISK based vaccine, and only suggested for horses that are at risk of contracting influenza due to their regular proximity to new horses (such as those that compete or race and are regularly exposed to other horses).

Rhinopneumonitis is caused by Equine Herpes Virus (EHV) of which there are three strains, EHV I, EHV III and EHV IV. It predominantly affects the respiratory tract (rhino = nasal, pneumo = lungs), but the EHVI and EHVIII strain may also affect reproduction causing abortion in mares. Symptoms of rhinopneumonitis include cough, fever and nasal discharge. Horses transmit EHV typically through horse to horse contact or via nasal or other secretions on objects (such as buckets). There is a RISK vaccine available for horses at risk of contracting rhinopneumonitis.

There is also a neurologic form of EHV1 infection called Equine Herpes virus Myeloencephalopathy (EHM), however this form is not prevented with the vaccine.

Strangles is also known as distemper and is caused by the *Streptococcus equi* bacteria. Symptoms of strangles include fever, nasal discharge and abscesses of the lymph nodes. It is highly contagious and the secretions from nasal discharge or abscess remain in the environment for a long time. There are two RISK vaccines available to prevent strangles, intramuscular and intranasal, though there is conflicting data available regarding their efficacy.

Equine Infectious Anemia is also called Swamp Fever. It is transmitted by vectors such as mosquitoes, deer flies or horse flies, or from mare to foal. It causes anemia, fever and edema. It is caused by a retrovirus, similar to what causes HIV in humans. There is no treatment for EIA and therefore horses that have EIA are either euthanized or donated for research to a private herd (that is isolated to prevent exposure to other horses). Carriers are identified through a Coggins test, which identifies if a horse has been exposed to the virus. There is no vaccine for prevention of EIA.

Potomac Horse Fever is also called equine monocytic ehrlichiosis and is caused by rickettsial bacteria. It is transmitted when horses consume insects (such as mayflies) that have consumed flukes infected with the bacterium , and is common in the Potomac region. Symptoms include fever, diarrhea and edema. There is a RISK vaccine available, but there are questions as to its efficacy.

Equine Viral Arteritis is caused by arterivirus and results in fever, depression, nasal discharge. It also causes abortion in mares or edema of the scrotum in stallions. It is similar to PRRSV (porcine reproductive and respiratory syndrome virus). It is transmitted through respiratory secretions and via sexual contact, and even through transported semen of infected stallions. Thus, all reproductive animals should be tested and cleaned prior to breeding, and should be vaccinated. Only those that are at RISK because they are breeding animals should be vaccinated.

Horses may also be vaccinated for some other diseases, such as botulism or some tick-borne diseases.

Section 9.3 Vaccinations and Immunity

https://aaep.org/document/adult-horse-vaccination-chart

Immunity refers to the ability to recognize foreign substances after an initial exposure. Foreign substances have antigens which allow the body to recognize them as foreign (non-self). The body responds by making antibodies to the antigen which helps identify the substances for attack by the immune system. Antibodies can be made in response to an initial exposure to the antigen that is natural (such as being exposed to chicken pox and contracting the disease, which results in antibodies being made to prevent subsequent disease) or through vaccines. Vaccines essentially train the body to recognize antigens of specific infectious agents.

Active Immunity refers to when the body ACTIVELY can make antibodies in response to an antigen. This requires a functioning immune system that is found in healthy mature horses. Passive immunity is when the body PASSIVELY receives antibodies such as when a foal consumes the first milk from the dam, called colostrum, or if an animal were to receive plasma via a transfusion.

Vaccines may be made from live modified viruses or bacteria, or from killed by intact viruses or bacteria. These allow the body to be exposed to the antigen so it can make antibodies against it, but the agents are modified enough (or killed) so the animal doesn't actually contract the disease. Tetanus vaccines are actually a toxoid against the toxin that causes tetanus (not the bacteria itself). An ideal vaccine provides strong prolonged protection and has no adverse side effects. Vaccines do not last forever, which is why an animal requires periodic boosters. Vaccines may be given intramuscularly (IM), under the skin (subcutaneously or Sub-Q) or intranasally via a mist spray.

The core vaccines that should be given to all horses include tetanus, eastern and western equine encephalomyelitis, rabies and west nile virus. Risk-based vaccines (those that are administered if the owner and veterinarian deems the animal at risk) include botulism, rhinopneumonitis, EVA, influenza, PHF and strangles.

Section 9. 4 Equine Parasites

https://aaep.org/document/internal-parasite-control-guidelines

Parasites are organisms that live off another animal, often at its host's expense. Parasites for horses are either internal (worms) or external. Symptoms of internal worms often are a result of the parasite using nutrients that were intended for the animal, resulting in a horse that is thin and has a poor hair coat. Because the parasite also often does damage to the horse's digestive tract or circulatory system, there may be other symptoms as well such as colic or internal bleeding. Most equine parasites have stages including eggs, several stages of larvae and adults. In many species, the eggs are hatched in the digestive tract and pass through into the manure. Eggs or larvae present on grass may then be inadvertently consumed by the horse.

Large Strongyles have a life cycle where the adults lay eggs in the horse's large intestine and are defecated in the feces, the larvae hatch and crawl up grass and are eaten by a horse. The larvae then can cause damage to the digestive tract, arteries and some organs before maturing to the adult worms. They are relatively large, 0.5-2 inches long. The S. vulgaris species migrates walls of arteries that supply the intestine, the S. edentatus migrates veins to the liver and the S. equinus penetrates the liver and pancreas.

Small Strongyles are also called cyathosomes and are most common in young horses. These worms stay in the digestive tract and cause damage to the mucosa, cause inflammation and anemia, and may cause colic. They can also become encysted and burrow into the intestines, which makes them challenging to treat.

Ascarids (Parascaris equorum) are roundworms that can become 6-12 inches in length. They typically only affect young horses. Horses consume the eggs, which hatch in small intestine, then the larvae penetrate wall and enter portal vein to the liver and then go to lungs where they are coughed up and reswallowed to become mature adults in the small intestine. Therefore they can damage the liver and the lungs, and the large adults can cause blockages in the intestine and colic.

Bots are considered both an internal and external parasite. The adult botfly lays eggs on the horse that look like little sesame seeds. The eggs hatch and the horse ingests the larvae. The larvae burrow in the mucosa of the stomach for approximately 10 months, then the bot larvae detach and pass out with the feces. They then become adult botflies that can go and lay 500 eggs. Bots can cause significant damage to the lining of the stomach.

Tapeworms involve an intermediate host, the mite. Horses eat the mite via grass or hay that is infected with the tapeworm. Tapeworms can rob the horse of nutrients.

Pinworms (Oxyuris equi) lay eggs near horse's anus, causes itching and the horse to rub its tail.

Parasitic prevention includes preventing horses from having access to any eggs or larvae. This can be done by keeping pastures clean by picking up manure or dragging manure to expose the eggs and larvae to sunlight (which kills them), and by rotating pastures with other animals (as most parasites don't cross contaminate). Also, one shouldn't overgraze pastures because that causes there to be less grass available for horses, making them more likely to consume grass that has been defecated on.

Owners and veterinarians can also conduct a fecal egg count test to determine how many eggs are "shed" by a horse. If a horse has less than 200 eggs per gram of feces, it is considered a light parasite load. However, not all parasites shed at the same time, and larvae are not counted in this test.

There are also medications that are dewormers called anthelmintics that kill parasites. Some strategies for deworming include either daily or continuous deworming, where a small dose of the medication is given to horses daily, or interval or strategic deworming, where medications are given at certain times of the year, aiming for specific parasites.

Some medications and brands are listed here.

> Daily **Pyrantel** (*Strongid*)– ascarids, strongyles (adults and larvae), tapeworms
>
> **Ivermectin** – ascarids, strongyles (adult and larvae), bots
>
> **Moxidectin** (*Quest*) – similar to ivermectin but good at getting the encysted larvae
>
> **Fenbendazole** (*Panacur*) – ascarids, strongyles (adult and larvae). "Power Pack" gets encysted larvae
>
> **Praziquantel** – tapeworms. Often given with ivermectin (*Equimax*)

External Parasites mostly annoy horses, but may also carry disease such as lyme disease, West Nile virus, EIA, etc. Examples of some external parasites include:

- Flies
- Gnats and midges
- Mosquitos
- Lice
- Fleas
- Mites and chiggers
- Ticks
- Ringworm

To control the exposure of external parasites to horses, one can control area where horses and decrease the parasites in the environment where horses are kept and to control access of the parasite to the horse (by using masks or blankets). Types of control to decrease the parasite population include insecticides, repellents, fungicides and growth regulators.

Further Reading

- Understanding Equine Preventative Medicine
  - Bentz, 2002
  - ISBN: 1581500866
- Equine Science
  - Pilliner and Davies, 2004
  - ISBN: 1405119446

CHAPTER 10

Equine Science

The objectives of this section are to learn about the basics of equine genetics, reproduction, nutrition and exercise physiology

Section 10.1 Equine Genetics

A **gene** is a hereditary unit of information found in a strand of **DNA** (deoxyribonucleic acid). Genes regulate various traits and functions within the body, such as coat color and blood type. The genotype is the genetic make up of a cell or animal, while the phenotype is the outward appearance of the animal, and is the result of the genotype plus the environment (for some traits; see heritability below). For example, the genotype would be the specific DNA code for a particular color, while "chestnut" would be a possible phenotype. DNA is found in all cells organized into chromosomes. Chromosomes are paired within cells, with horses having 32 pairs of chromosomes, thus having 64 total chromosomes within each cell. Thirty one of these pairs are considered homologous, meaning the pairs carry genes for the same trait, while the X and Y chromosomes are considered the sex chromosomes, code for different traits and in part determines the animal's gender (in addition to carrying other information). An animal with XX is female, and XY is male.

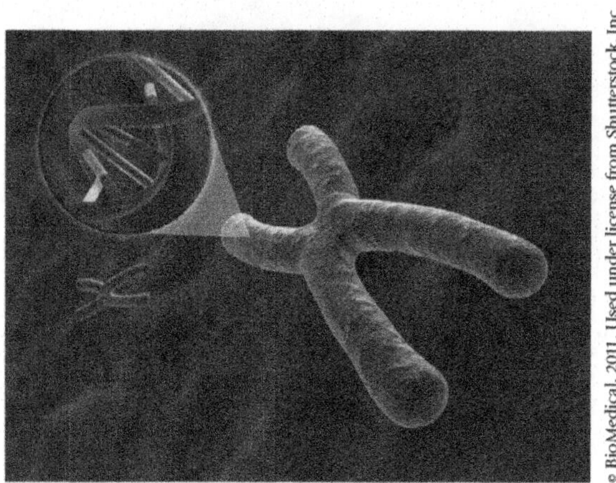

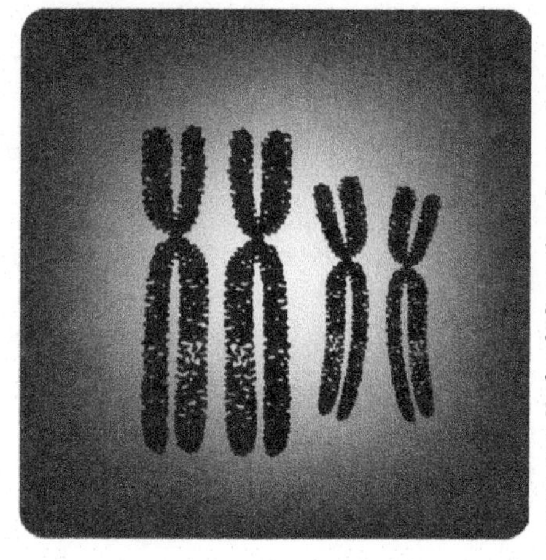

Of the pairs of chromosomes, one set of each pair comes from the sire, and one pair comes from the dam. During specialized cell division that occurs in gametes (eggs or sperm) called meiosis, each gamete only has one pair of chromosomes. (*Note this is different from normal cell division called mitosis which would create a duplicate daughter cell with 64 total chromosomes)

Thus, when an egg (32 chromosomes) is fertilized by a sperm (32 chromosomes), the pairs are reestablished and regenerates the 64 total chromosomes.

Species	# of Chromosomes	# Pairs
Przewalski's Horse	66	33
Horse	**64**	**32**
Donkey, Ass	62	31
Onager, Asiatic wild ass	56	28
Grevy's Zebra	46	23
Buchell's Zebra	44	22
Humans	46	23

Horse: 31 pairs of homologous (same) chromosomes plus 1 sex pair (XX or XY)

The term loci refers to the location or region of the gene on a chromosome, and may be used to describe that that gene controls.

An **allele** is a form of a gene that may result in a specific trait. Because each animal two sets of chromosomes from each of its parents, it has two alleles for each trait, and the combination of alleles may result in a specific phenotype. If both **alleles** are the same on the corresponding paired chromosomes, it is termed *homozygous*, while if the codes are different, they are *heterozygous*. Typically genetics uses letters to identify particular alleles, such as BB or bb (same letters, same alleles) to identify an animal that is homozygous for that trait, and Bb to identify an animal that is heterozygous for that trait.

The trait that is actually expressed depends on the *dominance* of the allele. If an allele is dominant, (often with a capital letter to designate its dominance) that will mask the other allele (where the other non-expressed allele is said to be recessive). Co-dominance is used when both alleles contribute to the phenotype. Epistasis is a term used when two or more genes interact, with some genes overriding others. This is very common in horse coat colors.

https://vgl.ucdavis.edu/resources/horse-coat-color

*Note this website above gives more details than required for this class!

The basic genetics of equine coat color relies on a few key principles; horses can only produce two main color pigments; eumelanin (causing black color) or pheomelanin (causing red color), or they can not produce any color at all (white). Where these colors (or lack thereof) are expressed on an animal gives rise to their overall color, and many different genes ultimately create a horse's coat color.

The base body color is caused by two key regions (loci); the Extension locus and the Agouti locus.

The extension locus is designated with the letter "E" and controls which color pigment is produced in a hair follicle. Eumelanin (black) is dominant (E) over pheomelanin (e). Thus, if an animal has a "big E" with either EE or Ee as its genotype, the animal will produce black hair. If an animal has ee as its genotype, it will produce red hair. If there are no other color modifiers (discussed momentarily) an animal with EE or Ee will be black, and an animal with ee will be red or chestnut.

    EE – black hair
    Ee – black hair
    ee - red hair

The agouti locus (A) is a modifying gene that controls **where on the body black pigment is** (therefore only of interest if horse is EE or Ee, because if they only produce red hair – ee – there is no black to restrict). The allele A restricts black to points (mane, tail, legs) only (dominant), while a results in no restriction. Thus an animal that produces black color (EE or Ee) AND has the genotype AA or Aa will be bay (black color is restricted to the points of the mane, tail and legs, leaving the body brown), while an animal with EE or Ee and aa will not have any restriction and will be black. (Again: if a horse is ee or red/chestnut, there is no black to restrict!).

All possible combinations of the E and A loci are:

EEAA – bay (black hair color, restriction of black to the points)

EeAA – bay (black hair color, restriction of black to the points)

eeAA – chestnut (red hair color, no black to restrict)

EEAa – bay (black hair color, restriction of black to the points)

EeAa – bay (black hair color, restriction of black to the points)

eeAa – chestnut (red hair color, no black to restrict)

EEaa – black (black hair, no restriction)

Eeaa – black (black hair, no restriction)

eeaa – chesnut (red hair color, no restriction/no black anyway)

Other genes also affect the outward color of the horse.

The W & G genes actually have the ability to override E and A.

The W gene results in white hair (no pigment) and is dominant, but it is lethal when homozygous dominant (WW). Thus, an animal that is Ww will have only white hair (no matter what other color genes it has), and an animal that is ww will have pigment (based on its other color genes).

The G gene results in the progressive whitening of hairs, resulting in a gray horse. It is dominant, so a horse with GG or Gg will be gray. The G gene is epistatic over other genes such as chestnut, bay or black.

The Cream or dillution gene (Ccr) causes a dilution or lightening effect of the base body color. If a horse has a base body color of bay, with one dilution gene the horse will be buckskin, or if it has two dilution genes it will be perlino. Similarly, a chestnut with one dilution is palomino (genetype eeCCcr), and two dilutions (double dilution; genotype eeCcrCcr) is cremello.

Main Color Genes in order of Epistatis

–W – white (W dominant over w) (WW does not exist in nature)

–G – gray (G dominant over g)

–E – extension red/black (E dominant over e)

–A – agouti black restriction (A dominant over a)

–C – cream (C co-dominant with Ccr to dilute)

–D – dun (D dominant over d)

–Plus many many others!

Some color examples

- WwGgEEAACCdd
  - White – W overrides everything after
- wwGgEEaaCcrCcrDd
  - Gray – G overrides everything after
- wwggeeAACCdd
  - Chestnut – ee=chestnut, no black to restrict to points
- wwggeeAaCCcrdd
  - Palomino – ee=chestnut, CCcr = single dilution

If you breed two palominos, what is the chance your foal is a palomino?

In many cases, one cannot tell the genotype simply by looking at the color. For example, a bay horse could be EEAA, EeAA, EeAa, or EEAa. But some color genotypes can be determined easily; for example a palomino is always eeCCcr. Thus, if one is breeding to achieve a specific color, genetic testing may be warranted.

There are several equine genetic diseases that are hereditary.

CID (Combined Immunodeficiency) or **SCID (S = severe)** is a condition that affects Arabians in which the immune system does not function properly. It is a homozygous recessive disease where animals with the genotype cc develop CIDS. Carriers (animals with the genotype Cc) should be identified and one should avoid breeding them. What is the chance of having a CID foal if you breed two carrier parents?

Hyperkalemic Periodic Paralysis (HYPP) or Impressive's Disease is a disease that results in a defect in the action of the sodium potassium ATPase pump. A famous Quarter Horse stallion (Impressive) had this disease and its prevalence increased due to his popularity as a breeding stallion. The disease results in muscle tremors, weakness, paralysis and potentially death.

–http://www.vgl.ucdavis.edu/services/hypp.php

It is an example of co-dominance, where an animal that has the genotype designated NN is normal, HH is severely affected by the disease, and animal with the genotype NH is moderately affected. The American Quarter Horse Association will not register horses with HH genotypes and is moving towards not registering those with NH genotypes in attempt to eradicate this disease in its breeding population.

Lethal White Foal Syndrome occurs with the overo gene that results in the overo pinto color pattern. Horses with this condition have a defect in their intestinal formation that is ultimately fatal. Horses with the genotype NN are normal colored (no overo coloring), horses with NO have overo color patterns, and horses with OO have the lethal white foal syndrome.

Other genetic conditions include hemophilia which is an x-linked condition (if an animal has a copy of the allele on its x gene) and results in a blood clotting disorder. Neonatal isoerythrolysis is a condition where a foal's red blood cells are destroyed by antibodies in the mare's colostrum. Polysaccharide storage myopathy (PSSM) is a condition common in Quarter Horses and draft breeds that results in a defect in their glycogen (stored form of glucose) storage.

Heritability (h^2) is the ratio of the genetic variance to the phenotypic variance. As indicated earlier, phenotype = genotype + **environment**. The extent to which a phenotype is influenced by the individuals genotype as opposed to environmental factors. It is specific to a population where $1 \geq h^2 \geq 0$ and when low (<0.2) environmental factors play a greater role.

Ranges for some traits

Trait	Heritability
TB race time	0.2-0.35
Trotting paces	0.2-0.35
Trotting earnings	0.2-0.35
Jumping, dressage	0.05-0.25
Conformation	0.3-0.5
Weight and height	0.5-0.9
Cannon bone circumference	0.4-0.8
Fertility	0.0-0.2

Genomics

Genomics is the area of study that encompasses examining all of the genes in an individual. The genome is an individual's entire set of DNA. Unlike genetics that focuses more on gene functions, genomics is the bigger picture. This is a rapidly growing area of biology – and goes far beyond what can be presented here.

The horse Twilight was a Thoroughbred mare, who was the first horse to have their entire DNA code sequenced. This was accomplished in 2009, and since then this information has been used to help better understand diseases, evolution and domestication.

A great summary can be found here: https://www.sciencedirect.com/science/article/abs/pii/S0737080621000861

Some highlights that have been discovered…

- Przewalski horses are actually feral, not wild
- Some horses have genetic traits that code for sprint (short) races, while others are more likely to be "stayers" (longer races).
- Some purebred Arabian horses are not actually purebred!

Section 10.2 – Reproduction

The vulva is the external part of the mare's genitalia. The vagina is also called the birth canal, and is where semen is deposited during mating. The uterus is where the fetus develops. The ovaries are where the eggs (ova) develop. The eggs develop in follicles throughout the ovarian cycle. The follicles are present since birth, but dominant follicles will develop due to the action of the hormone, follicle stimulating hormone (FSH) which is secreted from the pituitary gland. As the follicle grows and develops throughout the ovarian cycle, it produces estrogen, and it progresses into the Graafian follicle, which is a large follicle right before it releases the egg at ovulation. Ovulation is stimulated by luteinizing hormone (LH) which is also from

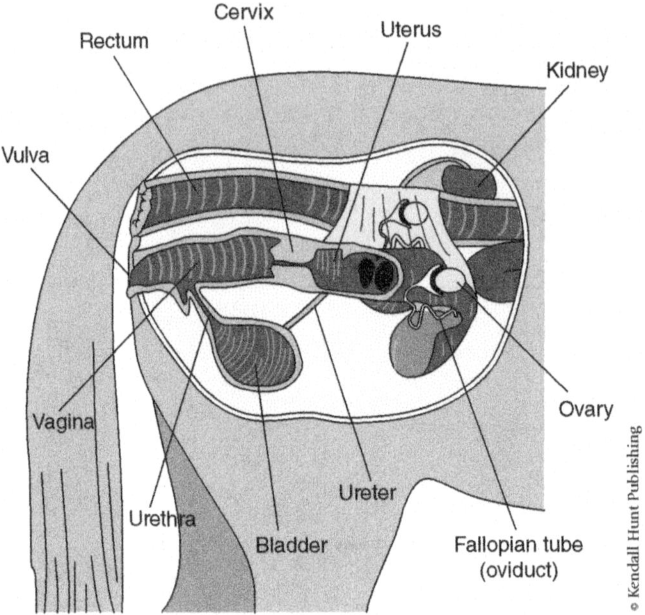

the pituitary gland. At ovulation, the egg is released into the oviduct (also called the fallopian tube) where it may be fertilized by sperm. If there is sperm present, the egg may be fertilized. If there is no sperm, the egg will dissolve and a new cycle will begin. After the egg is released from the follicle, the follicle becomes the corpus luteum, which secretes progesterone, a hormone important in preparing the uterus for implantation of the embryo (if one is present).

The estrous cycle refers to the changes of the ovaries in response to reproductive hormones (FSH, LH, estrogen and progesterone). The ovarian changes include follicular development (follicular phase also called proestrus), estrus (when the animal is sexually receptive or "in heat", occurs just prior to ovulation), ovulation, luteal phase (corpus luteum formation and subsequent regression, also called the luteal phase).

Horses are considered long day, seasonal breeders, and normally don't cycle in the winter months, a phase called anestrus. This is regulated by the hormone melatonin, which is produced by the pineal gland with darkness. Mares are often put under artificial lights, to help facilitate them to start cycling again earlier in the year.

There are other challenges to breeding in addition to mares being seasonal breeders. The egg, once released from the ovary typically only survives for 24 hours, which means that if fertilization is desired, there is a narrow window for breeding. Further, a stallion's sperm only lives for about 48 hours. Therefore it is very important to be able to determine exactly when a mare ovulates. Heat detection is often conducted by "teasing" a mare with a stallion, to determine if she is receptive or not. If she is in heat, she will display signs of heat such winking of her clitoris, and frequent urination. Alternatively, a veterinarian can conduct an ultrasound on the mare to determine the status of the follicle. Some mares will show a "silent heat" in which they will ovulate but not show any outward behavioral signs. This presents a further challenge for breeders.

If there is fertilization of the egg by a sperm, the resulting embryo will migrate throughout the uterus for approximately 2 weeks until it stays in one spot and then implants in the uterine wall (after about 40 days). The fetus develops within the **placenta** to supply it with nutrients, oxygen and remove wastes. The umbilical cord connects fetus to placenta to facilitate transfer of nutrients and wastes.

Pregnancy is detected through the use of ultrasound to detect an embryo or fetus. Ultrasound can also be done to ensure only one embryo is present (twinning is not desirable – why not?). A veterinarian can also palpate the uterus for changes that indicate a developing fetus. Equine chorionic gonadotropin hormone is a hormone that can be detected with a blood test to also determine if a mare is pregnant.

Gestation (the period when the fetus develops) is approximately 335 days. There are three trimesters; the first trimester where there is slow growth of the fetus, but significant development of the placenta, the second trimester where there is rapid growth of the fetus, and the third trimester which also has significant growth. At birth, the foal will weigh about 10% of the mare's weight.

Parturition is the term for foaling or giving birth. Prior to parturition, the mare will show signs that she is close to giving birth, such as waxing of her teats, enlargement of her milk veins and colicky behavior. There are also kits available to detect the calcium and pH of any dripping milk that may indicate an impeding birth.

There are three stages of labor. In the 1st stage the foal moves into position until water breaks and this stage may last several hours. In the 2nd stage the fetus moves through the birth canal and is born. There are intense uterine contractions and this stage happens very quickly. The 3rd stage is the expulsion of the placenta. It is very important to inspect the placenta for any tears, and to ensure it is expelled rapidly (typically after about 30 minutes). If any of the placenta is retained within the mare it is considered a veterinary emergency. Dystocia is the term for trouble foaling, and may be a result of improper position of the foal.

The newborn foal should stand and nurse within 60 min of birth. The foal should nurse to ensure it gets colostrum (the first milk, which is rich in antibodies), and the foal should be assessed to ensure passive transfer of antibodies was sufficient. The umbilical cord should break on its own and disinfectant should be applied to the stump (chlorohexidine or dilute iodine). The foal should pass the meconium (first feces) within 3 hours after birth.

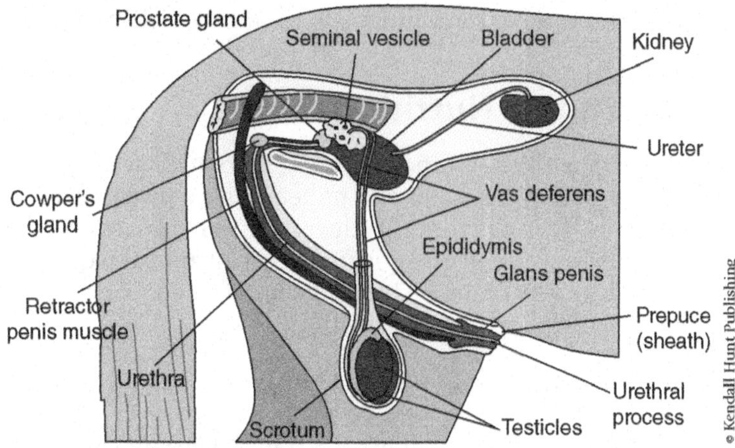

In the stallion, the testicles (testes) produce both sperm and testosterone. They develop inside the body until just before birth, and then they descend into the scrotum. If they do not descend and one or both of the testes remain in the body, the horse is termed a ridgling (cryptorchid). The testes are often removed at a young age, resulting in a gelding.

http://www.acvs.org/AnimalOwners/HealthConditions/LargeAnimalEquineTopics/
UndescendedTesticlesinHorses/

Sperm produced in the testes are stored in the epididymis. During mating the stallion will get an erection, and with ejaculation, the sperm will travel up the vas deferens and into the urethra. The Cowper's Gland (also called the Bulbourethral gland), seminal vesicle and prostate gland produce fluids and nutrients that are added to the sperm, resulting in semen.

There are several methods used to breed horses. "Live cover" refers to when a stallion mounts a mare, and ejaculates directly into the mare's vagina. Artificial insemination entails the sperm being placed in the mare by a veterinarian or technician.

Pasture mating involves allowing a stallion to have free access to a mare (or several) in a large area. The stallion is usually very good at detecting heat, and can usually get a high percentage of the mares pregnant. However, it is unknown when conception took place. Further, there may be injuries to the mare or stallion during mating.

Hand mating entails leading the stallion to a mare. This procedure has the advantages that one knows exactly when the mare was bred, and precautions can be taken to prevent injury to the mare or stallion. However, it requires intensive management (to determine when the mare is in heat) and may be dangerous to the handler (s).

Artificial insemination has two steps. First, semen is collected from a stallion using a "phantom" dummy horse to mount, and the semen is collected into an artificial vagina. The semen quality is inspected and doses are calculated and aliquots are made and shipped. The second step involves the veterinarian placing the semen into the mare's vagina using a syringe or "straw". Some breed registries do not allow this procedure.

Embryo transfer is a process by which a mare is impregnated (either through live cover or artificial insemination), but then the embryo is removed (flushed out) and placed into a recipient mare. The fetus grows in the recipient mare, but has the genetics of the donor mare and the stallion. The benefit however is that the donor mare can now cycle again, and perhaps become pregnant with another foal (thus allowing more than one foal per year) or can continue to compete in athletic events.

Section 10.3 Equine Nutrition

The nutrients required by the horse include; water, carbohydrates, fat, protein, vitamins and minerals. Carbohydrates, fat and protein can be metabolized to provide fuel (energy) for the body in the form of adenosine triphosphate (ATP). ATP is required for most cellular functions. The quantities that an animal requires of each of the nutrients depends largely on its body weight, but also on its physiologic status, such as lactation, pregnancy, growth or work.

Digestion

In order for the horse to obtain these nutrients, it must consume feeds that contain them, and break them down into small enough particles that they can absorb and use. Digestion is the process by which feed is broken down through the actions of chewing, enzymes or microbes, into smaller particles. These smaller particles can then be absorbed from the digestive tract into the bloodstream. Once nutrients are into the bloodsteam, they may be metabolized and used for various functions.

Nutrients and particles that are not absorbed remain in the digestive tract and become feces.

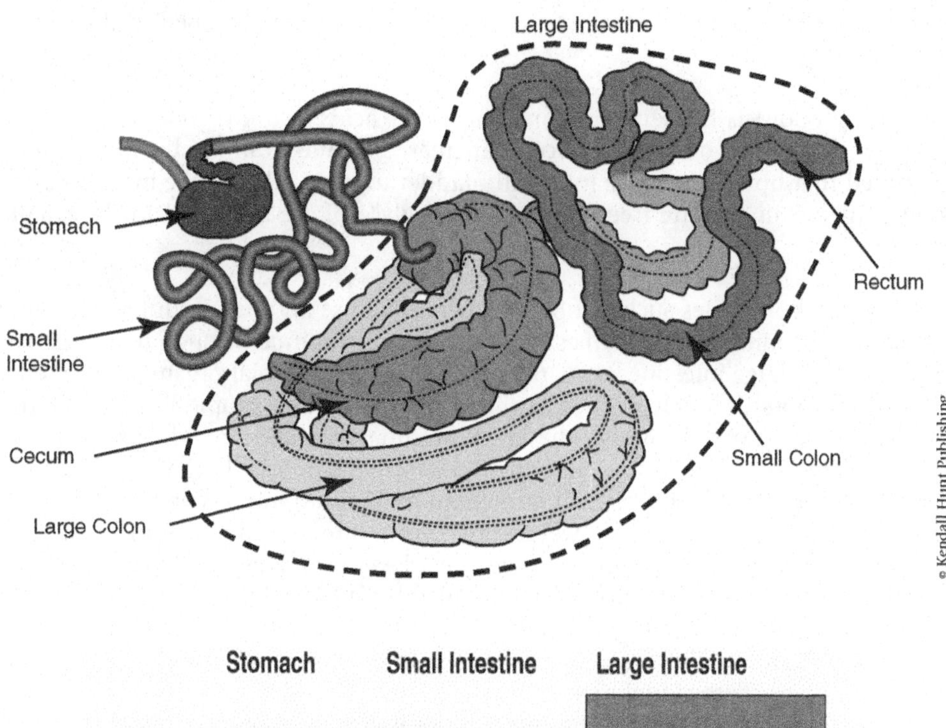

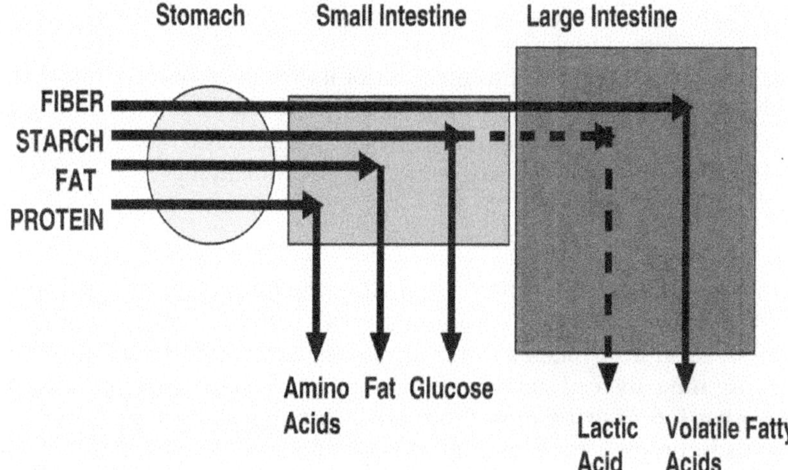

The gastrointestinal tract (GIT) of the horse is complex. Horses use both enzymatic and microbial digestion to break down feed particles. Enzymatic digestion occurs in the "foregut" (stomach and small intestine), while microbial fermentation occurs in the hindgut (cecum and colon).

Foods are consumed by the horse, who uses its teeth and lips to grasp food particles. Feeds are chewed and saliva released from salivary glands to help lubricate the food for swallowing. Food travels down the esophagus to the stomach, where it mixes with hydrochloric acid and some enzymes. From there food continues through the small intestine (the parts include the duodenum, jejunum and ileum) and into the large intestine (the cecum, large and small colon). In the large intestine there is a massive population of microbial organisms (bacteria, protozoa, fungi) that assist digestion through fermentation. Any food not being digested and absorbed continues out the rectum into the feces.

Here is a good video: https://www.youtube.com/watch?v=maWXVKI-gq4

Protein is broken down beginning in the stomach through the actions of the hydrochloric acid, and then broken down further by enzymes in stomach and small intestines. The individual amino acids are released and are absorbed in the small intestine.

Fat (triglyceride) is broken down into free fatty acids and glycerol components through the actions of enzymes from pancreas, and bile from the liver. It should be noted that horses do not have a gall bladder to store bile, unlike humans and most other animals. The free fatty acids and glycerol are absorbed in the small intestine.

Complex carbohydrates such as fiber (namely cellulose and hemicellulose) are indigestible by mammalian enzymes. Thus, these will continue through the digestive tract until the large intestine. Here, the population of microbial organisms ferment fibrous compounds to release the volatile fatty acids, acetate, propionate and butyrate. These are absorbed through the large intestinal wall and provide a significant source of energy for the horse. All herbivores form a similar relationship with microbial organisms to facilitate fiber fermentation to allow animals to consume and use plants as food. The horse however differs from humans, in which there is little microbial fermentation (as omnivores they eat less fiber), and is different from ruminants where the fermentation occurs in the foregut (rumen).

Simple sugars and starches are primarily broken down in the small intestine to glucose. Glucose is then absorbed in the small intestine. If a horse consumes large quantities of starch, the horse may not be able to break all of it down before it continues through the digestive tract to the large intestine. In the large intestine the microbial organisms ferment starches into volatile fatty acids and lactic acid. Fermentation of starch also produces significant quantities of gas, which can be problematic for the horse.

Water

Water is the most important nutrient required by the horse. Horses will drink an average of 10 gallons per day, but this amount varies with body size and other things. Feed type can affect how much a horse drinks, such that a horse will drink more if they are consuming dry feed such as hay, but will drink less if they are eating pasture that already has lots of water in it. If a mare is lactating she will have higher water requirements to produce milk, and an athletic horse will need additional water to replace any lost in sweat. Horses will also drink more or less based on the environmental temperature. If a horse does not consume enough water they may be at risk for impaction or blockage of the digest tract (due to digesta being too dry to move through the digestive tract easily). Also, if they don't drink enough the horse may become dehydrated.

Energy/Calories

Energy (measured in **calories**) is required for all processes in the body. *Energetic substrates* that can be broken down and metabolized for energy include carbohydrates, fat and protein. Carbohydrate sources include simple carbohydrates such as basic sugars (glucose) and starches (long chains of glucose) that come from feeds such as corn, oats and barley; and complex carbohydrates are fibers (longer complex chains of glucose such as hemicellulose or cellulose) that come from forages such as hay or pasture. Fat is an excellent source of energy for the horse, providing more than 2.5 times the calories per unit weight than carbohydrates. Sources of fat include vegetable oil, fish oil and other feeds such as rice bran that are high in fat. Protein can be metabolized and used for energy, however it is a very inefficient process. Horses require different amounts of energy depending on their physiologic needs and body size; with athletic horses and lactating mares having very high requirements.

As indicated, energy is measured in calories. The amount of energy in one calorie is enough heat to heat and raise 1 gram of water by one degree centigrade. Clearly this is a small amount, so most human nutrition resources use "Calorie" (with a capital C) which is actually 1 kilocalorie (1,000 calories). In horse nutrition, we use the term "Megacalories" (or MCal) which are 1000 kilocalories/Calories or 1,000,000 calories.

Calories (in terms of heat) are measured in feeds through bomb calorimetry, which really measures the heat produced when a feed is set on fire, and this is called "gross energy". Carbohydrates have 4 mcal/kg, fat has 9 mcal/kg and protein has about 5 mcal/kg. When feeds are consumed, they are usually made up of some protein, some carbohydrates and some fat – so feeds might have a gross energy value of 4.6 mcal/kg, for example, due to the different energy contributions from fat/starch/protein in a feed.

When a horse consumes feeds, the gross energy in the feed is not really all available to the animal. A large portion of feed may be indigestible, and winds up in the feces, so the energy available to the horse is really the "digestible energy". So while things like hay and oats might have similar amounts of carbohydrates and similar GROSS energy values – they would have very different digestible energy values because so much energy is lost during the digestion/fermentation of hay compared to oats.

A 500 kilogram horse at maintenance requires 16.7 Mcal of digestible energy per day, or a sample of hay might have 1.9 Mcal of digestible energy per kilogram.

Horses do not have an actual daily requirement of carbohydrates, such as they do not need a certain amount of grams of carbohydrates per day. This is because carbohydrates are really only present in the diet as an energetic substrate. However, fiber, while not really needed for metabolism, is essential to keep the microbes of the digestive tract happy… and it is often said that if a horse's microbial environment isn't healthy – then the horse won't be happy. So while one might talk about feeding their horse a "low carbohydrate" or "low carb" diet – they are really only trying to reduce the simple carbohydrates – starches and sugars – in a horse's diet. We would never want to limit the amount of fiber in a horse's diet. Some horses do need to limit the amount of simple or soluble carbohydrates (starches and sugars) due to metabolic disease, muscular diseases or behavior (some horses may get more hyper after eating a lot of sugar, similar to some kids!).

Similarly, the major function of "fat" is to be metabolized for calories for horses. However – there is some evidence that some types of fats may be better than others, and horses may have some requirements for these. Fats are triglycerides – so they have 3 fatty acid chains attached to a glycerol unit. Fats differ based on the number of carbon units they have in their fatty acid chain, as well as the number of bonds (single bonds or double bonds) they have between some of those carbons. Saturated fats are "saturated with hydrogen" – and so all of the carbons are connected by a single bond, with the extra points attached to hydrogen. Saturated fats are usually solid at room temperature. Unsaturated fats have carbons not attached to as many hydrogens, and instead have double bonds between carbon units. Polyunsaturated fats have multiple (poly) double bonds. These double bonds usually happen at the third or sixth spot from the "omega" end of the fatty acid chain – giving rise to omega-3 or omega-6 fatty acids. The body can't insert a double bond into the 3rd position – so horses need to eat some omega-3 fats. These are plentiful in pasture, but are lower in grains. Also – there are some health benefits associated with omega-3 fats – like some anti-inflammatory properties and some immune benefits. Great sources of these "good" fats are fish oil, and to a lesser degree other oils like linseed, camelina, etc.

Protein

Protein is a chain of building blocks called amino acids, and is a component of body structure such as muscle and connective tissue, and is a component of enzymes and hormones. Amino acids are compounds that include Nitrogen (unlike most other compounds within the body) in their basic core structure, and then a unique side chain to give rise to the 20 or so different amino acids. Amino acids also can be broken down and used for energy, but this typically only happens with endurance types of exercise or starvation, when other fuels are depleted, or may happen if too much protein is consumed. Of the different amino acids, some are able to be synthesized by the body, while others cannot. These are considered the "essential" amino acids, because they are dietarily essential and required to be in the diet. Different feeds have different quantities of these amino acids, therefore it is important to not only consider the total protein quantity of a feed, but also its protein quality (referring to the range of essential amino acids within it). One of the most limiting amino acids is lysine. Good protein sources include legume hays (such as alfalfa or clover) and seed meals such as soybean meal or linseed (flax seed) meal.

Thoughts…

> Which diet would provide more protein?
> 500 grams of a feed with 20% protein?
> 1 kg of a feed with 10% protein?

*They both provide the same amount of protein! Don't focus as much on % of protein on feed bags… focus on how much protein a horse might get from eating a reasonable amount of it.

Also – a horse could eat 3000 grams of protein (WAY more than any horse would need) – but if they didn't have lysine (an essential amino acid) then the horse would suffer from a nutrient deficiency.

And if a horse ate 3 times as much protein (and amino acids) than they need – they can't store it for later use. Instead – that extra protein is broken down, and ultimately winds up as urea and stinks up your horse's stall! Protein is also often the most expensive component of your horse's diet, so there is no need to overfeed. In fact – too much protein may be detrimental to your horse's health and performance.

Vitamins

Vitamins are nutrients that function to support metabolic processes, usually through supporting enzyme activity as "cofactors". The vitamins are categorized as either water soluble (vitamin C and the B complex) or fat soluble vitamins (A, D, E and K), depending on their nature in water.

Vitamin C can be synthesized from glucose in the liver of horses (unlike humans, guinea pigs and non-human primates), therefore it is not required in their diet. Some horses may however benefit from additional vitamin C, such as older horses (that may not be able to produce vitamin C as well) or athletic horses (as vitamin C is an antioxidant and may help repair exercise-induced tissue damage). The B complex of vitamins includes niacin, thiamin, riboflavin, folic acid, cobalamin and biotin, and these are involved in metabolism. These vitamins are found in adequate amounts in most horse feeds, and they are also synthesized by the microbes found in the horse's digestive tract. Therefore, horses typically do not have a problem meeting their B vitamin needs. However, research has shown that biotin may be beneficial to horses with poor hoof quality, and therefore biotin supplementation may be warranted.

Vitamin A is required for vision and the support of tissue growth. Horses can consume beta-carotene, which is found in leafy forages and carrots, and then converted to vitamin A. Vitamin D is the "sunshine" vitamin, and is synthesized by the skin upon exposure to sunlight, and is also found in sun-cured forages such as hay. Vitamin D functions support calcium metabolism. If horses get adequate exposure to sunlight they generally don't need vitamin D in their diets. Vitamin E functions as an antioxidant and is found in adequate amounts in most horse feeds, but in high amounts in fresh forages such as pasture. If a horse only consumes stored forages (such as hay) they may need vitamin E supplementation. Vitamin K is required for blood coagulation, and while it is found in most forages, it also can be synthesized by microbes found in the horse's digestive tract.

So – between the horse's typical diet (pasture) and their healthy digestive tract (microbes) and some sunlight – most horses don't need a whole lot of vitamin supplementation. However – if they don't get out to pasture much, eat old hay, and are blanketed a lot, they might need some supplementation – particularly for vitamins A and E (and D?).

Minerals

The minerals are categorized as either macro (major) or micro (minor) minerals, depending on the quantity of their needs. Calcium and phosphorus are the major components of bone, while calcium is also required for muscle contraction, and phosphorus is required for energy metabolism (as the phosphorus part of ATP). Sodium, chloride and potassium are considered electrolytes, and are required to maintain cellular electrical activity and to help regulate hydration and blood pH. Other macro minerals include magnesium and sulfur. The microminerals include copper, zinc, iron, iodine, manganese and selenium, as well as many others.

Most feeds are rich in many minerals that a horse needs, with the exception of sodium. Some other minerals – zinc, copper, selenium – may also need to be supplemented.

Feeding Horses

To feed a horse to provide it with its nutrient requirements, the diet must provide adequate levels of the nutrients, be palatable and not have any toxins. Also, the diet is ideally economical to the horse owner.

The MAJOR component of a horse's diet (other than water) is **forage**, such as hay or pasture. The horse's complex digestive tract was designed to deal with a large forage intake, and gastrointestinal health requires sufficient fiber to help keep the microbial population stable, and health consequences can occur if a horse does not consume enough forage. A rule of thumb, is that horses should consume AT LEAST 1% of its body weight as dry forage. In addition, forages can provide significant quantities of all of the required nutrients, in particular if it is high quality, such as pasture. The grazing behavior associated with forage consumption is also important to the horse's time management, and horses that do not consume enough forage often develop behavioral conditions such as weaving, wood chewing or cribbing, which will be discussed later.

If a horse is consuming adequate forage, it is likely he comes close to meeting his nutrient needs. However, if a horse cannot consume enough forage (there is an upper limit to how much they can consume) or forage quality is low, they may need to be fed additional feeds to meet their needs. Such feeds may include concentrated energy feeds such as cereal grains, by-products or fat sources, or higher protein feeds such as soybean meal. Horses may also need to be supplemented with additional vitamins or minerals, based on the nutrient composition of the forage, or any additional feeds they consume.

A horse at "maintenance" however (one not growing, lactating, working, etc) has relatively low requirements, and can meet all of his nutrient requirements through consuming water, forage (hay or pasture) and a salt source (as most forages and other horse feeds tend to be low in sodium).

Forages refer to the leafy stalks of plants – and typically includes pasture (fresh plant), hay (dried plant) or haylage (partially dried, preserved plant). Grass hay is hay made from dried grasses, such as timothy, orchardgrass, brome, fescue, bluegrass etc. Grass hays tend to be lower in protein and calcium than legume hays. Legume hays are those made from legumes, such as alfalfa, clover, birdsfoot trefoil, and are higher in protein and calcium, though tend to be more expensive. Often hay is a mix of grass and legume. While some information about the hay's nutrient quality can be assessed by knowing what plants are in it and gauging its level of maturity, the only way to know the exact nutrient composition of your hay is to have it analyzed.

Hay Analysis

Forage Form: Hay   **Ext. Assistance:** N   **Production Status:** N/A
Species: Horse   **Maturity:** N/A
Forage Type: Mixed Mostly Grass   **Special Treatment:**

Species	Analyte	Unit	As Submitted Basis	Dry Matter Basis
	Dry Matter	%	89.75	
	Crude Protein	%	16.03	17.86
	Unavailable Protein	%	1.70	1.89
	Adjusted Crude Protein	%	15.93	17.75
	Nitrate Ion	%	0.15	0.17
	Neutral Detergent Fiber	%	47.01	52.38
	Acid Detergent Fiber	%	32.50	36.21
(Horse)	Digestive Energy	Mcal/lb	0.94	1.05
(Horse)	Non-fiber Carbohydrate	%	19.61	21.85
	Fat	%	1.72	1.92
	Calcium	%	0.71	0.79
	Phosphorus	%	0.46	0.51
	Sulfur	%	0.17	0.19
	Magnesium	%	0.65	0.73
	Sodium	%	0.01	0.02
	Potassium	%	0.91	1.01
	Copper	ppm	11.00	12.00
	Iron	ppm	122.00	136.00
	Manganese	ppm	31.00	34.00
	Zinc	ppm	32.00	36.00
	Ash	%	5.38	5.99

Pasture plants also includes grasses and legumes, though the grasses may be further categorized as warm season grasses (Bermuda grass, bahia grass) or cool season grasses (timothy, bromegrass, fescue, etc).

As indicated, forages can provide significant quantities of most nutrients. The following graph shows how well major nutrients are provided for, in different types of horses – assuming a 500 kg mature body weight and consuming 1.75% of their body weight as forage.

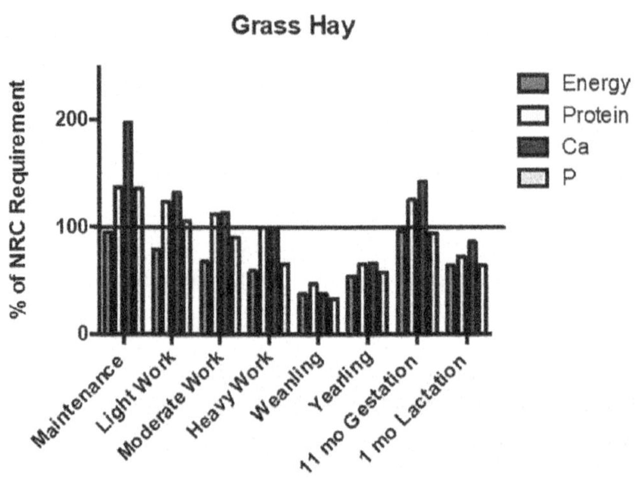

If energy needs are not met through forage consumption, a horse may need to consume concentrated energy feeds, such as whole grains. These include oats, corn, rye, barley or wheat, and they are good sources of calories but are low in protein (and lysine) and calcium. By product feeds may also be fed as energy sources, and have variable amounts of protein and minerals. These include wheat bran, wheat middlings, beet pulp and rice bran. Beet pulp and rice bran are also good sources of fiber.

Energy Feeds

Feed	Energy	Protein	Ca	P
Oats	2.85	11.8	0.08	0.34
Corn	3.38	9.1	0.05	0.27
Barley	3.26	11.7	0.05	0.38
Beet Pulp	2.33	8.9	0.62	0.09
Wheat Bran	2.94	15.4	0.13	1.13
Vegetable Oil	~9	—	—	—

Because of the highly variable amounts of nutrients found in "raw" ingredients such as cereal grains or by-product feeds, horse owners often turn to feed companies to provide feed mixes that are formulated for specific classes of horses. These commercially available feeds are classified based on how the ingredients are mixed; textured feeds (also called sweet feeds) are those where the raw ingredients are simply mixed and one can identify all of the particles within them. Pelleted feeds have been ground together and put through a pelleter, while extrude feeds are ground and processed with heat and pressure to produce a particle similar to puffed dog food. All of these types can be fortified with different amounts of energy, protein, vitamins and minerals mixed in quantities suitable for an intended horse class. Examination of the feed tag can give information about the crude protein, crude fiber, fat, sodium, calcium, phosphorus, salt and selenium content. Some tags may also provide information about other nutrients, feed instructions and ingredients. It is important that horse owners never feed commercial mixes formulated for other species.

Any type of concentrate (grain, commercial feed, etc) should only be offered to horses if the nutrients required by the animal are not provided for in sufficient amounts from the forage (when requirements are increased due to growth, work etc), if good quality forage is not available, or to modify behavior (such as to help you catch your horse!). Horses are often fed too many concentrates and there are increasing numbers of obese horses as a consequence.

Commercial feeds are designed for different types of horses – such as for growing horses (which would be fortified with extra nutrients needed for growth) or performance horses (typically higher in calories). Many horses do not need more calories because they can eat enough hay or pasture. However, they may still need some nutrients. If they only need some vitamins and minerals, they could be offered a small amount of a highly concentrated vitamin-mineral mix (usually fed at a rate of less than 115 grams/ 4 oz per day). Alternatively, horses may need a bit of protein as well (if the hay/pasture quality isn't as good) – and they may be fed a Balancer or Ration Balancer. These products contain high amounts of protein (often at 30%) but are fed at a lower rate (~500g – 1 kg; 1-2 lbs per day).

For example, a commercial feed for a performance horse may have 4.5 mcal/kg and 14% protein. That performance horse might eat 3 kg of the feed, and get 13.5 mcal of energy and 630 grams of protein from it. For a horse with lower calorie needs (ie. Not in extensive work, and/or able to meet calorie requirements from forages alone) – they might consume a ration balancer with 3.5 mcal/kg and 30% protein, but might only consume 1 kg of it. Then that horse would get 3.5 mcal of energy, and 300 grams of protein – which would be more appropriate for its requirements.

*Remember – the percentage of protein on a tag only matters when you know how much of it you are feeding. Weigh your feed!

There are several non-nutrient ingredients available as supplements for horses that may be considered herbs or other functional foods. Some of these have some decent research behind their benefits, while others do not. Do some research before spending a lot of money on these! Also be wary that different supplements may contain a common ingredient (for example, vitamin E or selenium) and if you feed multiple supplements you might be feeding potentially toxic levels.

Nutrition Related Disorders

Colic is a term used to describe digestive upset in the horse. It is a consequence of gastrointestinal problems such as impactions (blockages), or twists, often as a result of the complexity of the horse's digestive system (with many turns and changes in diameter). Colic can be caused by non-nutritional factors, such as parasite infestation, but may also be caused by diet. In particular, a disruption to the microbial population within the large intestine can result in excess lactic acid or gas production, and subsequent death of some organisms (which can produce toxins). Sudden changes in diet and high concentrate/low forage diets are two key nutritional causes of colic. Horses fed in sandy areas may also be at risk for sand build up and subsequent blockages. In general, horses should be fed as little concentrate as possible (where >5kg of grain per day increases the risk of colic by 6 times!) and any dietary changes (to grain or forage type) should be introduced over several (10+) days.

Gastric ulcers are another digestive condition that is relatively common in horses. Normally there is a balance between the acid production of the stomach with the protective secretions to buffer and protect the underlying tissues. When there is an imbalance, the acidity of the stomach can damage the stomach tissue, resulting in an ulcer. Providing as much forage as possible to the horse is one way to prevent gastric ulcers, as calcium found in forages (in particular legumes such as alfalfa) can buffer acidity. Further, the saliva produced through chewing has natural buffers to help counteract acidity. Having long stretches with no food increases the acidity of the stomach and puts a horse at risk for ulceration. Ulcers can be treated medically, but are far cheaper to prevent where possible through good feeding management.

Obesity is often associated with overeating. Body weight can be managed by eliminating excess calorie intake (decreasing/removing grain, replacing with small amounts of highly concentrated feeds such as balancers or supplements), and increasing exercise.

Laminitis can be associated with obesity, but may also be triggered by feed events – such as a horse eating too much starch from getting into a feed bag, or even excessive pasture intake.

Botulism is rare in adult horses, but may be acquired if a horse consumes haylage that was not made properly.

There are several types of poisonous plants for horses. In many cases, horses will not eat poisonous plants unless they have nothing else to eat. However – it is wise to be aware of what poisonous plants live in your area and clear them from your pastures. It is also possible that some common feeds can be infected with toxins. It is always wise to purchase feed – hay and concentrates from the best possible sources.

Section 10.4 Exercise Physiology

Skeletal Muscle is made up of muscle fibers, which are muscle cells. They can range in length, from a few millimeters to several centimeters long. Parts of a muscle fiber (cell) include the sarcolemma (the outer cell membrane), multiple nuclei, glycogen granules (the storage form of glucose within the cell), fat droplets, myoglobin (which can store oxygen) and mitochondria (the energy producing organelles). Muscle also contains contractile proteins called actin and myosin that facilitate the shortening (contraction) of muscle.

When horses (and people!) start to exercise, many internal signals tell the heart rate to increase, respiratory rate to increase, blood flow to muscles to increase… ultimately to deliver more oxygen to the muscles. To a large degree – we can measure the intensity of a work load by measuring (or estimating) oxygen consumption, or by monitoring heart rate. Both will increase during increasing efforts of exercise, and will eventually plateau. With respect to oxygen uptake, this volume of oxygen when it plateaus is called the Maximum Volume of Oxygen Update – or VO_2Max. While heart rate max doesn't change as much with fitness – the VO_2Max is highly adaptable with exercise conditioning – and can be used to track exercise training efforts, and to compare athletic abilities between athletes – and between species!

For example, an untrained human might have a VO_2Max of 50 mls of oxygen per kilogram body weight per minute, a trained human might be closer to 80-90 ml/kg/min.

An untrained horse might be close to 120 mls/kg/min, while a trained horse might be close to 200!

Similarly, the speed/work intensity that you might reach your VO_2Max might increase with training.

It should be noted that it is possible to work or run (a little) harder than your VO_2Max – you would just be doing work without oxygen! (see below!)

Muscle contraction uses energy in the form of ATP, and if a muscle is going to contract and relax multiple times (as with exercise), the body needs to replenish the muscle with ATP. ATP can be generated through both aerobic (with oxygen) and anaerobic (without oxygen) metabolism. Anaerobic metabolism is generally used in short bursts of exercise where ATP production needs are short but needed rapidly, often before sufficient oxygen is available from increased respiratory rates, such as at the start of exercise, or in an intense activity, such as a Quarter Horse race. The pathways of anaerobic metabolism include the creatine phosphate system and anaerobic glycolysis (the breakdown of carbohydrate such as glucose or glycogen to lactic acid). Aerobic metabolism is used when there is adequate oxygen available, and the exercise intensity is relatively low. Substrates for aerobic energy production include carbohydrate (glycogen and glucose), fat and protein, and results in the production of carbon dioxide. Again, we don't want to use protein for energy production because the excess nitrogen following its metabolism winds up on the stall floor!

There are different types of muscle fibers, generally categorized as either fast twitch (fast contracting) and slow twitch (slow contracting), though there are several subtypes. Fast twitch muscle fibers are those associated with rapid contraction, and have high glycolytic activity (to facilitate anaerobic metabolism) and fatigue easily. Slow twitch muscle fibers are those that make use of aerobic or oxidative metabolism and have low fatigability. Horses that are better at endurance sports (like Arabians) tend to have more slow twitch types of fibers, while those breeds that excel at speed (like Quarter Horses) tend to have more fast-twitch fibers.

With many types of exercise, a horse will fatigue and not be able to continue work at a given intensity. With high intensity exercise (such as with a Quarter horse or Thoroughbred race), fatigue is often caused lactic acid buildup in the muscles. In lower intensity exercise such as an endurance race, fatigue is often a result of the horse running out of fuel reserves, such as muscle glycogen.

Exercise training is useful to help develop endurance so a horse doesn't fatigue, and to help develop muscle and tendon strength. With training, the body becomes more efficient at a given level of work. For example, at a given speed, the heart rate will be lower because the heart is more efficient at pumping blood, and the tissues are more efficient at extracting and using oxygen. Horses also become better able to regulate body temperature through sweating. Overall, there is also a shift for more fat oxidation with aerobic work, to help preserve carbohydrate (glycogen) stores.

In general, horses are excellent athletes. They have a very large cardiac output, such that the amount of blood pumped per minute (a function of both heart rate and the amount pumped in each contraction) is high. This is a result of the very high heart rate horses are able to achieve with exercise (240+ beats per minute, compared to approximately 30 beats per minute at rest). Horses also have a high maximum oxygen uptake, which means that their tissues are able to extract and use large amounts of oxygen to help sustain ATP generation. Horses are very good at sweating, which allows them to dissipate heat easily. Horses also have the ability to release red blood cells from the spleen, which allows them to carry more oxygen in their blood.

Cardiac Output Values

	Rest	Exercise
Human	5 l/min	35 l/min
Horse	25 l/min	300 l/min

Per kg

Human	0.35 l/kg/min
Horse	0.7 l/kg/min

Oxygen Uptake

- Horse
  Up to 180-200+ ml/kg/min
- Human
  65-85 ml/kg/min

Despite these features of athleticism, horses do have challenges. They can develop low blood oxygen (hypoxemia) during exercise, perhaps due to an inability to increase respiratory rate sufficiently (because it is related to stride). Horses can also become dehydrated easily because they sweat so profusely. They are also prone to musculoskeletal issues due to the extreme weight and concussion placed on their limbs.

Some conditions that are related to exercise include EIPH (discussed earlier), PSSM (polysaccharide storage myopathy; a glycogen storage disorder), recurrent exertional rhabdomyolysis (RER; a condition seen in Thoroughbreds and Standardbreds that results in muscle damage) and anhidrosis (where a horse cannot sweat, resulting in an inability to dissipate heat and regulate body temperature).

Further Reading

- Equine Color Genetics
  - Sponenberg, 2009
  - ISBN: 0813813646
- Equine Breeding Management and Artificial Insemination
  - Samper, 2008
  - ISBN: 9781416052340
- Equine Clinical Nutrition
  - Remillard, 2023
  - ISBN 978-1-119-30369-5
- The Athletic horse
  - Hodgson, McKeever, McGown, 2014
  - ISBN: 978-0-7216-0075-8

CHAPTER 11

Equine Behavior and Learning

The objectives of this section are to learn about the horse's natural behavior, how they learn and about abnormal behaviors

Section 11.1 Normal Equine Behavior

Some behaviors a horse displays are considered instinctive, that do not need to be taught, and are partly based on genetics. Such behaviors include standing, suckling, running and vocalizations. Horses are a precocious species, meaning they are relatively able to fend for themselves upon birth (compared to for example kittens or even human babies). Other behaviors are learned as a result of responses and consequences to the environment and events, and these take longer to develop.

There are some key characteristics of equine behavior that should be understood in effort to understand how a horse learns. Horses are prey animals, and therefore have a fight or flight instinct, with flight being the most common response when threatened. Horses are herd animals and therefore develop a pecking order to establish dominance amongst the heard. This type of relationship can also be established with other animals, including humans. Horses can communicate through many mediums, including vocalization and body language. These can be interpreted to some degree by people and used with training. In addition, horses are very perceptive, and can interpret cues given to them relatively easily, even subtle cues such as a shift in body weight. Horses are relatively fast learners and have an excellent memory, which facilitates their ability to be trained.

Horse are reactive, and will respond to stimuli accordingly. For example, when it rains or is windy and cold, they will seek shelter. Eating behavior is centered around grazing, where horses will spend upwards of 18 hours per day grazing, thought the time spent depends in part based on season, weather and vegetative availability. Food selection is based on palatability, and horses are usually very good at avoiding toxic plants (unless little other food is available). Horses also tend to avoid areas that have been urinated or defecated on, which often results in pastures with "roughs and lawns" where there are closely grazed areas and those with long grass that is avoided. Some horses will display distinct eliminative behavior, urinating and defecating as a means to mark territory. Drinking is based on availability, and horses may dig to find water when scarce. Sexual behavior involves mating and courtship, but also refers to the mare's relationship with her foal. Horses are herd animals and will give and seek care from others. Mutual grooming is a behavior where horses will stand head to tail with another horse and groom each other, usually around the withers or shoulder. This also helps decrease flies around their faces, as one horse's tail swishes near the other's face. Horses will show aggression and fight, though usually this is limited to stallions establishing their dominance. Horses like to mimic another, and if one horse shies or runs, others tend to as well, in a "stampede" effect. This can be useful for training purposes, when horses can be introduced to new situations by following another horse (such as leading them through water, etc.). Horses tend to be curious and offering them an opportunity to inspect a new situation is helpful. Horses have the ability to rest while standing thanks to their stay apparatus, which can help wild animals be ready to run away quickly if threatened. Horses do need recumbent (lying down) sleep as well, and if in a herd will take turns, with the horse(s) sleeping in the middle, while others will stand on lookout. Horses will spend 20-30% of their day resting.

© JustASC, 2011. Used under license from Shutterstock, Inc.

In the wild (or in feral populations), horses will organize into bands, or groups. A harem band is a group that is lead by the harem stallion, and includes several mature mares and immature offspring. Bachelor bands are a group of non-breeding juveniles that have yet to form their own harem. Alternatively a lone stallion may roam. There are several advantages to forming bands; such as increased survival thanks to more eyes to watch for predators and the ability to defend resources. Large bands also allow for genetic diversity to continue. However, injuries from group stampedes are possible, and in times where resources are limited, sustaining a large group may be difficult. Further, infectious disease may affect an entire band.

Photos courtesy of author

Horses are social animals and are able to communicate effectively among individuals and across groups. Horses can communicate using a variety of mediums, including visual, auditory, chemical and tactile means. Visual communication includes facial expressions, body, head and ear positions. Aggression is easily recognized, with horses showing signs such as ears pinned, stomping feet, swishing their tail or barring their teeth. Acoustic signals include neighs and whinnies, squeals, snorts, groans and roars. Horses may also stomp their hooves as an auditory signal. Chemical signals are a result of chemicals called pheromones, that are found in the skin, saliva, urine and feces. These chemicals are involved in communication, identification and sexual courtship. Tactile communication includes calming signals such as mutual grooming and a mother guiding her foal to nurse, or aggressive signals such as biting.

Domesticated horses will exhibit many of the same types of behaviors as wild or feral horses, though they may be limited by housing and management. Horses kept at pasture in large groups often will organize into bands. Stabled horses will not manage their own time, but it will be based on feeding and/or turnout schedules. Breeding domesticated horses is obviously very different from in wild or feral populations. Not only is the relationship between the mare and stallion different, but the mare and foal's relationship will differ, as foals are often weaned much earlier in managed situations. Often, because of management schedules, horses are at risk for developing unwanted behaviors,

Section 11.2 Equine Learning and Training

Learning can be described as a modification of behavior, or the ability to perform a behavior differently as a result of experience. Alternatively it may be defined as gaining knowledge. Horses tend to learn as a result of being exposed to different stimuli (triggers that cause a behavior). Unconditioned stimuli are those that cause a response with no training or practice, and may be considered reflexive. For example, if you were to make a loud noise, your horse may move away from the noise. A conditioned stimuli is one where the response has been learned through practice, and may be considered a cue, such as applying pressure with the leg to encourage the horse to move in a particular direction or gait. Habituation refers to a decrease in the response to a stimulus (desensitization), such as an animal getting used to clippers. This concept is often encouraged in younger horses, through Imprint Training (discussed more later) or "sacking out". Sensitization however is when a horse becomes more responsive to a stimulus as a result of experience and learning, such as a dressage horse performing movements based on very subtle cues.

There are two main theories of learning. Classical conditioning was theorized by Pavlov, and Operant Conditioning was theorized by Skinner. There are several other theories as well. Classical conditioning results when an established behavior occurs in response to a new stimuli. Pavlov used dogs, and had an unconditioned stimulus (food) and an unconditioned response (salivation). Pavlov trained the dogs by ringing a bell (conditioned stimulus) whenever the dogs were fed. Eventually, the dogs would salivate (conditioned response) when the bell was rung. This type of learning is common in horses, who might start to whinny in anticipation of food any time the feed door is opened.

Operant conditioning is based on the premise where a response that is rewarded is more likely to occur again. Skinner used rats (and other animals) and placed them in a box with a lever, and if they hit the lever they got a reward such as food. While initially they may have randomly hit the level, eventually the animals would learn to hit the lever to get the treat. This theory thus is based on reinforcement, or an event following a behavior. An appetitive reinforcement is one that increases the likelihood of a behavior, while an aversive reinforcement decreases the likelihood of a behavior. Types of appetitive reinforcement includes positive (+) reinforcement where something is given (added) in response to a desired behavior. An example would be giving your horse a treat when they do something you want – like get onto a trailer. Negative (-) reinforcement is based on a stimulus being removed (taken away) when a behavior is done. An example of this would be the use of a rump rope to teach a horse to walk in hand. If the horse does not move forward, pressure is applied at their buttock, but if they move forward the pressure is removed. The same premise lies in the use of choke or prong collars in dogs, or lip chains in horses, where if the animal behaves and keeps their head in the correct position and doesn't pull there is no pressure, but if they pull there is pressure (and if they stop pulling the escape the pressure). Aversive reinforcement aims to decrease an undesirable behavior. Positive (+) aversive reinforcement refers to punishment, when something is applied/added (such as a bite or whip) in an attempt to decrease a behavior. While punishment can be effective, it does not train the horse to do the correct action, and may result in aversion to the handler or even decreased effectiveness. Negative (-) aversive reinforcement refers to removing something in attempt to decrease a behavior. This is difficult to accomplish in horses, but is very effective in children – such as taking away a toy if they are misbehaving!

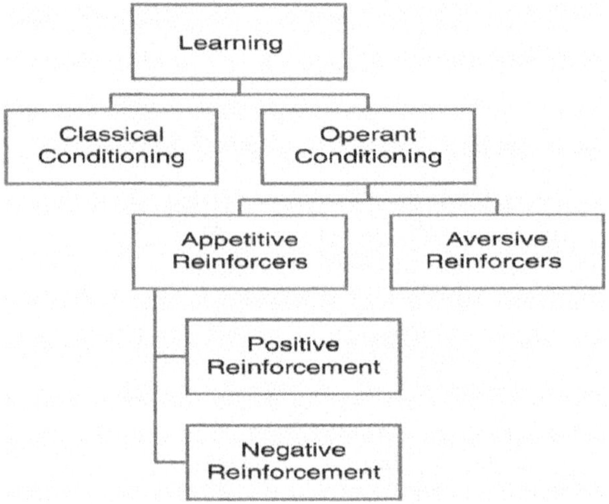

Training horses can use all of these principles, but in general positive reinforcement tends to be the most effective. Conditioning your horse to be satiated with secondary reinforcers such as a pat or voice (rather than a treat) is helpful. It should also be noted that the rate of learning varies, in part based on the ability of the trainer, but also the horse's preparedness (both mentally and physically).

Imprint training is a common training technique for young horses that aims to desensitize them early on in life. Some believe that the first 48-96 hours of life have a unique window of opportunity for enhanced learning. Getting a foal used to human interaction and handling in this time is common, and facility managers will often work extensively with young animals. While some handling is warranted and effective, some research suggests that foals may be more aversive towards people with intensive imprint training.

Section 11. 3 Undesirable Behaviors

Undesirable behaviors of horses include actions such as biting, shying, bucking, rearing and stereotypic behaviors known as stable vices. In many cases, these behaviors are the result of either time management issues, improper training, or even pain (such as bucking due to an ill-fitted saddle).

Stereotypies are repetitive, abnormal behaviors, often called stable vices. Up to 34% of all domestic horses show some kind of stereotypic behavior, and these are not seen in the wild. They tend to be caused by frustration or boredom, lack of forage or grazing time, or lack of socialization. Some are based on confinement (stall weaving, pawing or stall walking) while others are oral (cribbing and wood chewing).

Weaving is when a horse will continuously rock his head and body from side to side. It can put wear on the front feet and shoes and puts strain on the front legs, and result in weight loss. Solutions include giving the horse more forage (hay), a companion animal or putting up a yoke gate. Environmental enrichments such as toys, stall balls and mirrors may help. Pawing is a behavior that is common in the wild (to dig for water, to find food under snow and for communication), but in a stable setting it is often a result of boredom and a desire for attention which is often rewarded (by the owner saying "stop pawing"). Stall walking is when a horse will walk in circles in their stall, resulting in a messy stall and weight loss. Mirrors have been shown to decrease the behavior by up to 97%.

Cribbing is a behavior where a horse places his incisors on a solid surface, pulls back and sucks in air. Some horses are able to wind suck without using an object. This behavior can be detrimental by potentially causing colic or digestive problems, and wears down the edges of the incisors. There are "cribbing collars" that decrease the frequency of the behavior, but must be worn regularly. Wood chewing tends to be caused by a lack of fiber in the horse's diet as a result of too little forage being available. Electric fences and topical applicants can decrease this behavior.

© Thomas Barrat, 2011. Used under license from Shutterstock, Inc.

In general, dealing with stereotypic behaviors involves decreasing the cause of the problem (boredom) and increasing the amount of forages available to the horse, ideally through grazing at pasture (which may also provide socialization). Environmental enrichments such as toys and mirrors are also very effective at decreasing stereotypic behavior.

Further Reading

- The Ultimate Horse Behavior and Training Book
  - Tellington-Jones and Lieberman, 2006
  - ISBN: 1570763208
- Understanding your Horse: How to overcome common behavioral problems
  - Bayley and Maxwell, 1997
  - 9781570760730

CHAPTER 12

Equine Management

The objectives of this section are to learn about how to manage horses, equine buildings and the costs associated with horse ownership

Section 12. 1 Daily Care of Horses

Managing an equine facility involves daily care of horses, as well as overall record keeping and management. Day to day activities include things like feeding, turnout to pasture and cleaning of the stalls. Preventative health care includes vaccinations, parasite management and farrier care. Individual horses may have specific needs, such as blankets or leg boots, etc. Business management includes record keeping such as horse owner information (emergency contact information, insurance), health records, etc. There are several commercial products available to help with record keeping.

Good horse identification is useful not only for health records (Coggins testing, passports), but also to help with recovery from theft or natural disaster. Methods of identification include pictures or drawings of colors and markings, lip tattoos, freeze brands and microchips.

Daily care includes feeding and watering, but the frequency of such may depend on the facilities. For example the use of automated waterers simply requires an insurance that they are working. Automatic feeders may also be useful in some facilities. How hay might be provided to horses in a paddock or pasture (if pasture quality is low) is an important consideration. Large round bales of hay are usually cost effective, though wastage of hay (due to stomping on it or rain) can be high. Hay mangers, hay huts or other devices to prevent wastage typically pay for themselves over time. Stall cleaning should be done at least daily, and many facilities will clean the stall several times per day. Horse grooming may be done daily, or at least a whole-body look over to check for cuts or injuries.

Young horses have special needs, in particular a foal. Weaning is a management strategy that is beneficial (yet highly stressful) to both the mare and foal, and the age when a foal is weaned depends largely on the management practices. There are a few methods for weaning that are used, including "cold turkey" where the mare is removed abruptly, a physical separation where the mare and foal can see each other but the foal can't nurse, timed separation where the mare and foal are separated for increasing amounts of time per day, or a single removal weaning system, where within a group of mares and foals, one mare is removed at a time. Young horses also need intensive training, feeding protocols and exercise to enhance bone development.

Business management involves keeping good financial records to record things like income and expenditures. Some expenditures are fixed (such as a mortgage), while others are variable (such as vet bills). A facility should have insurance, such as Commercial Equine Liability insurance or Care, Custody and Control insurance.

It is good practice for a boarding facility to have a boarding contract with horse owners. This should include the barn rules and requirements prior to the horse arriving, such has having a negative Coggins test, emergency information and insurance information. Properties should also display the Equine Liability Act sign, that limits a stable's liability in case of injury.

The Equine Liability Sign states:

WARNING.

Under North Carolina Law, an equine activity sponsor or equine professional shall not be liable for an injury to or the death of a participant in equine activities resulting exclusively from the inherent risks of equine activities.

Chapter 99E of the North Carolina General Statutes

Section 12. 1 Buildings and Equipment

The facility itself includes the building, land, fences and equipment (farm implements, barn supplies, tack, etc.).

The barn itself should be designed to be convenient and safe, but should also take into considerations for the horse as well, such as facilitating socialization. Ventilation (though windows, cupolas or forced air), lighting and heating (if needed) should be considered. The facility should be located on level ground, with drainage away from the facility, and following all local zoning regulations. Most barns are built from either metal, brick or wood, with advantages and disadvantages to each.

Photos courtesy of author

Facilities should be designed with a manure management plan. Horses can produce up to 50 lbs of manure per day, plus 6-10 gallons of urine. When added to bedding, this can amount to up to 100 lbs per day of waste. Options for manure management include removal, spreading the waste on fields, or composting.

Fire safety is very important in a barn, where things such as bedding and hay are highly flammable. Fire detection is very important, and barns should be equipped with ample fire extinguishers and ideally a sprinkler system. Lightning rods can also help prevent fires from lightning strike ignition. Designing the barn to help facilitate escape routes can be useful. Rules and signs about not-smoking is very important in the prevention of fires.

The stalls themselves should be approximately 12 x 12 (though 10 x 10 is a minimum), with larger stalls (12 x 18) for mares and foals. Flooring type is an important decision with barn design, with concrete, dirt, and rubber mats as options (which may in turn affect bedding selection). Bedding can be shavings, dust, straw, peat moss, or other novel products, and selection may depend on cost and convenience.

Fences should be 48 tall to prevent escape, with double fences around the perimeter and stallion paddocks. Types of fencing includes PVC, post-and-board, woven wire, Diamond V, high tensile wire, elastic rope, or ribbon and pipes. The type chosen depends on costs and maintenance, as some (like wood) often requires more maintenance than others (pipe or PVC). Barbed wire should never be used for horses.

Photos courtesy of author

Pastures should be managed to maintain the nutritional quality of the forages present and minimize the parasite population. Dragging pastures can break up manure piles to expose parasitic eggs to sunlight, while rotational grazing helps the pasture plants re-establish themselves to provide sufficient growth.

It is important that all horse facilities be prepared for emergencies, such as hurricanes, tornados, flooding, and fire. Planning ahead involves making sure there is always a sufficient supply of feed and water, as well as measures to recover horses (such as thorough identification) is important.

Section 12.3 Horse Ownership

The purchase of a horse is the cheapest part of owning a horse.

There are many websites available for buying and selling horses.

- www.equine.com
- www.horsefinders.com
- www.bigeq.com

When purchasing a horse it is important to remember that the buyer has the control over the deal. A pre-purchase exam is when a veterinarian comes and does a check on the horse (at different levels of thoroughness, based on costs) to determine if there are any preexisting conditions. It may be worthwhile to have a trainer come and ride the horse or have a trial period prior to purchase.

Depending on your own situation, it may be necessary to board a horse.

Full board is when the facility provides all feed and services and the horse has its own stall and turnout area, and may cost between $350 to $2,000 and up per month. Pasture board is when horses are kept at pasture only (no stall) and may cost between $150 to $300 and up per month. Co-ops are becoming more popular and ranges from $200 to 400 per month. All costs are based on location and facilities.

Alternatively it may be possible to keep your horse on your own property at home. While significantly cheaper than boarding typical costs include:

- Hay ($5 to $15 per 50 lb bale, 2 to 5 bales per week)
- Grain ($15 to $40+ per 50 lb bag)
- Bedding
- Labor (your time is money!)
- Facility maintenance and upkeep

Other costs associated with horse ownership include farrier care ($40 to $100+ per trim—reset 8 times per year $320 to $800 per year), preventative health care (vaccines, deworming, dental care at $200 to $400 per year) and health insurance. Equine mortality insurance is like life insurance for your horse while equine major medical is health insurance for your horse. The costs are highly variable and depend largely on the value of your horse ($50 to $500/year).

Other extra costs of horse ownership include things like lessons, showing, transportation, tack, grooming supplies, blankets, etc. Potential owners should also consider how long you plan on owning your horse. Would you agree to selling your horse some day? Or would you keep him until he dies? (which may be well into the horse's 30s.)

The cost of owning a boarded horse: 10-year-old gelding who lives to be 30 years of age

- Boarded for ~$500/month for 20 years $120,000

Plus any veterinary, supplement, tack/equipment, insurance horse shows/travel, etc.

The cost of owning a horse kept at home: 10 year old gelding, Lived to be 30 years…

- Kept at home
- Hay: $600 per year
- Grain: $300 per year
- Farrier: $600 per year
- Bedding/Manure maintenance: $800 per year
- Vet: $300 per year
- Overhead: $380 per year
- Facility Maintenance: $500 per year

TOTAL ~$69,600 over 20 years (still a lot of money)

*There are ways to keep costs to a minimum – but the bottom line is that horses are expensive.

Section 12.4 Equine Careers

The best way to make a million dollars in the horse industry is to start with 2 million!

There are several different career options for those who love horses. Many equine students aim to own and manage a boarding or training facility. Many decisions come with such a career, such as how much to charge per month, will you do all of the labor or hire someone, will you offer training or lessons, etc. Other careers may be those that have daily hands-on activities with horses (such as a massage therapist or veterinarian) or those that are supportive of the horse industry (managing a tack or feed store, running horse shows, marketing companies, journalism, selling veterinary pharmaceuticals, education, law, etc.). Career options also depend on the type of education you might be willing to obtain.

A veterinarian is a very popular career choice for horse lovers who are committed to many years of university. Depending on your goals, career types of veterinarians may include working as a clinician, working in academia (working and teaching at a university), or working in industry (for example, at a company that makes vaccinations or other drugs). You may choose a specialty based on your interests, such as internal medicine, surgery, ophthalmology, pathology, dentistry, lameness, reproduction, etc. Schooling to become a veterinarian usually includes 3 to 4 years of undergraduate university education plus 4 years of veterinary school. Speciality areas as described earlier may also require 4-10 years of additional training and schooling. Admission to veterinary school is highly competitive, and once you are there, it is very expensive (with student loans often outweighing earning potential). There are also some negative aspects to the career itself, with death and neglect a part of the job. Prospective veterinarians should also consider if they want to work in a private practice or a clinic, and should consider the positive and negative aspects of each. Of course, the rewards of being able to spend the day with horses and save lives may outweigh any negative concerns.

Earning a graduate degree in equine science is another academic path students should consider. After an undergraduate degree, a Master of Science degree is typically 2 years, and a doctoral degree (PhD) is typically 3 to 4 years. The career outcomes for an equine scientist depend on their specialty (reproduction, nutrition, genetics, exercise physiology, cell biology, etc.) and their goals. Academia (teaching, research, and extension) is a popular option, while working in industry (such as at a feed company) often has higher earning potential. If you attend graduate school and work on a research thesis and serve as a teaching or research assistant, you may also be able to obtain funding that pays for your education as well as a small salary. This may be a more desirable avenue to enter a higher paying career (similar if not more than veterinarians) without having to pay for student loans.

A good thing to do as you look at different careers is to talk to as many people as possible with different jobs, ask to shadow different careers or find internships or even employment opportunities in those areas. You might not know you like something until you try it, or you might have your heart set on a career but realize you don't like it as much once you try it. Get out there and make connections!

Other things to consider as you look into equine careers…

- Where do you want to live? In a big city vs. in a rural area. Which jobs are where?
- Do you want a family? Some careers are more supportive of families than others (sadly)
- Do you also want horses for your hobby? Sometimes combining your work and your hobby can make your hobby feel like work.

- How much money do you want to make? Money isn't everything, but it can be important. Consider looking up salaries and salary-debt ratios, living expenses in your desired area – housing, schooling, daycare, etc.
- How much time at university or other education do you want? Sometimes it can be hard to see other friends go and have disposable incomes while you're busy at school.
- Consider "outside the box" – there will be jobs in 5-10 years that aren't even a thing yet!

GLOSSARY OF TERMS

- **Allele:** Form of the gene (code)
- **Anthelmintic:** The proper term for deworming (antiparasitic) medication
- **Arthritis:** Inflammation of the joints.
- **Bay:** A horse with red or brown hair on its body and black "points"—its mane, tail, and lower legs
- **Bench Knees:** Cannons fail to come from the center of the knees (front view)
- **Black:** A horse with black hair on its body and a black mane and tail
- **Blaze:** A facial marking with a wide stripe down the horse's face
- **Body Condition Score:** A score (usually on a 1–9 scale) assigned to a horse based on how much fat coverage it has (1 = emaciated, 9 = grossly obese)
- **Bow Knees:** Horses stand over the outside of their front feet. Can cause side bones (front view)
- **Bowlegged:** When a horse stands pigeon-toed on his hind feet with hocks turned out
- **Breed:** A group of horses with common ancestry and/or common characteristics
- **Bridle:** A piece of tack that is used for riding; includes a *bit* that goes in the horse's mouth and *reins* that are held by a rider
- **Broke:** A horse that is well trained (broke to ride)
- **Buck Knees/Knee Sprung:** Over at the knees (side view)
- **Buckskin:** A horse with a yellow or beige body color with black points
- **Calf Knees:** Knees that break backwards (side view)
- **Camped Out:** Rear legs set out behind the hip
- **Camped Out/Under:** Legs out/under vertical line (side view)
- **Camped Under:** Rear legs set too far under
- **Canine (teeth):** Usually only found in males, also called tushes
- **Chestnut/Sorrel:** A red or brown body color with brown (or red or flaxen) mane and tail
- **Coggins Test:** A test used to identify horses that have been exposed to the virus causing Equine Infectious Anemia
- **Colt:** A young male horse
- **Concentrate:** A term used to describe feeds (grain mixes) for horses with concentrated energy (calories)
- **Coprophagy:** When a horse (or other animal) eats feces
- **Core vaccine:** a vaccine that all horses should get yearly (eg. Rabies, tetanus, EEE/WEE, EWNV)
- **Coronet Marking:** A leg marking with the white just around the coronet band
- **Cow Hocked:** Horse stands with point of hocks turned inward, base wide and splay footed
- **Cremello:** A horse that is light colored over its body as a result of a double dilution of a chestnut horse
- **Cross-Firing:** When the rear foot strikes the front foot on the opposite side of the body

- **Curb Bit:** A type of bit in which the reins attach to shanks to use leverage
- **Curry Comb:** A grooming tool used to lift dirt up from the horse's skin
- **Dam:** A horse's female parent
- **Dandy Brush:** A grooming tool used to remove dirt from the horse's hair
- **Deciduous Teeth:** Baby teeth
- **Dental Star:** Dark line and eventually spot made from the pulp cavity
- **Dilution:** When a gene (often called the dilution or cream gene) dilutes (lightens) the color of a horse.
- **Draft:** A large build type of horse, weighing upwards of 1400 lbs
- **Dun:** A horse with a yellow or beige body, dark mane and tail and "primitive" markings such as a dorsal stripe or zebra stripes on its legs
- **Embryo transfer:** A method to produce multiple offspring from one mare in a year, or to be able to keep a mare in work, while getting an offspring from her. A mare (donor mare) is impregnated, and then the embryo is removed (flushed) and implanted in another mare (recipient mare). The donor mare can then be bred again (and again and again?) or could return to work.
- **Epistasis:** When two different genes interact
- **Equitation:** Refers to riding horses. Equitation competition classes would be judged on the rider's ability.
- **Farrier:** The person who looks after the horse's hooves. They may also be a blacksmith – which refers to a person who can use heat and metal to create horse shoes. Some farriers would only purchase ready-made shoes.
- **Feral horse:** A wild (free ranging) horse that is a descendent of a domesticated horse
- **Fetlock Marking:** A leg marking with the white going up over the fetlock joint
- **Filly:** A young female horse
- **Floating Teeth:** A term used to describe when a veterinarian or equine dentist files the teeth to remove the sharp points
- **Foal:** A newborn horse
- **Forage:** General term for the plants that horse (and other livestock) eat – primarily grasses and legumes, as fresh pasture, or processed as hay, haylage or other hay product (cubes or pellets)
- **Forging:** When the rear foot strikes the front foot on the same side of the body
- **Furlong:** A unit of measure for horse racing, where one furlong is 1/8 of a mile, or 220 yards.
- **Gait:** The order of footfalls when the horse is in motion. Includes the walk, trot, canter and gallop plus other breed specific gaits
- **Galvayne's Groove:** A groove that appears on the upper corner incisor in older horses
- **Gamete:** Egg and sperm cells
- **Gelding:** A mature male horse that has been castrated (or "gelded")
- **Gene:** Unit of information from DNA
- **Genotype:** The genetic makeup of an animal – usually referring to the different alleles an animal may have
- **Genome:** The complete set of genetic material in a cell
- **Gray (color):** A horse with a mix of white and dark hairs over black skin
- **Green:** A term used for a horse that is "unbroken" or newly in training
- **Hackamore:** A bitless bridle

- **Halter:** A piece of tack that goes on the horse's head and is used for leading or tying
- **Hand:** Term for measuring horse height: 1 hand = 4 inches
- **Heritability:** The ratio of the genetic variance to the phenotypic variance
- **Heterozygous:** Both codes or alleles are different on the pair of chromosomes
- **Hinny:** The cross between a stallion and a jennet (donkey)
- **Homologous Chromosome:** Pairs of chromosomes with the same basic traits
- **Homozygous:** Both codes or alleles are the same on the pairs of chromosomes
- **Hoof:** the horse's "feet" – but more accurately their toes - including the hoof wall (equivalent to fingernail)
- **Hoof Pick:** A grooming tool used to remove dirt and stones from the horse's hooves
- **Horse:** A hooved mammal of the Equidae family (*Equus ferus caballus*)
- **Incisors:** Teeth, central, middle and corner
- **Infundibulum:** Also called the cup. Dark (due to food build up) depression on the grinding surface of teeth. Leaves as an enamel spot
- **Knock-knees:** Horse stands in at the knees or is too close at the knees (front view)
- **Lead Shank (or lead rope):** A rope or leather "leash" for leading horses
- **Light Horse:** A small framed horse
- **Longeing (lunging) (a horse):** Having the horse on a very long line (a "longe line") going around in a circle with the handler in the middle
- **Mare:** A mature female horse
- **Meiosis:** Specialized cell division of gametes
- **Molars and Premolars:** Back teeth suited for grinding
- **Mule:** The cross between a mare and a jack (donkey)
- **Overo:** A type of pinto pattern
- **Paint:** A breed of horse that shows pinto coloring
- **Palomino:** A horse with a golden or yellow color on its body with a light mane and tail
- **Pari-Mutuel Betting:** A type of betting system where a fixed percent of the wager goes to the track
- **Pastern Marking:** A leg marking with white going up over the pastern joint
- **Perlino:** A horse color that is the result of a double dilution of bay
- **Permanent Teeth:** Adult teeth
- **Phenotype:** The outward appearance of an animal (based on genetics *and* environment)
- **Pigeon-toed:** Toed in (front view)
- **Pinto:** A horse that is a mix of blocks of white with any other color
- **Pony:** A horse with small mature size
- **Post-legged:** Too little angulation in the hock
- **Ration Balancer:** a type of feed with highly concentrated nutrients (eg. Protein ~ 30%) that can be fed in small amounts to limit calorie intake from the feed
- **Roan:** A color of horse where the white hairs are mixed with the base body color to give an overall lightened effect
- **Roman Nose:** A type of head where the nose is convex

- **Sabino:** A type of white spotting pattern that often results in lots of white
- **Saddle:** A piece of tack that is placed on the horses back and where the rider sits for riding
- **Scalping:** When the front foot hits the hind foot at the hoof on the same side of the body
- **Sickle Hocked:** When horse's rear legs have too much angle at the hock and resemble a sickle
- **Sire:** A horse's male parent
- **Snaffle bit:** A type of bit in which the reins attach directly to the piece that goes into the horse's mouth
- **Snip:** A facial marking with a spot between the horse's nostrils
- **Sock:** A leg marking with white going up half of the cannon bone
- **Splay footed:** Toed out (front view)
- **Stallion:** A mature male horse
- **Star:** A facial marking with a small spot between the horse's eyes
- **Stay Apparatus:** The mechanism by which horses can rest while standing
- **Stimuli:** A trigger that causes a behavior. With training, these may be considered "cues"
- **Stocking:** A leg marking with white going up the entire cannon bone (to knee or hock)
- **Stripe:** A facial marking with a narrow stripe down the horse's faces
- **Tack:** A general term for the equipment used on the horse
- **Tobiano:** A type of pinto pattern
- **Tovero:** A type of pinto pattern
- **Warmblood:** A medium framed horse often originating from Europe
- **Weanling:** A young horse that has been weaned from its dam
- **Yearling:** A horse that is one year of age (up to 2 years of age).
- **Zoonotic:** A disease that can be spread between humans and animals (eg. Rabies)